Performance Assessment for Process Monitoring and Fault Detection Methods

Kai Zhang

Performance Assessment for Process Monitoring and Fault Detection Methods

Springer Vieweg

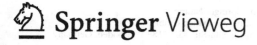

Kai Zhang
Duisburg, Germany

Dissertation, Duisburg-Essen University, 2016

ISBN 978-3-658-15970-2 ISBN 978-3-658-15971-9 (eBook)
DOI 10.1007/978-3-658-15971-9

Library of Congress Control Number: 2016954529

Springer Vieweg

Printed on acid-free paper

This Springer Vieweg imprint is published by Springer Nature
The registered company is Springer Fachmedien Wiesbaden GmbH
The registered company address is: Abraham-Lincoln-Str. 46, 65189 Wiesbaden, Germany

To my parents and Sissi

Preface

With the increasing demands on product quality and process operating safety, process monitoring and fault detection (PM-FD) has become an important area of research in recent decades. Numerous methods were developed in this area for different types of processes and applied to various industrial sectors. However, there is little work focusing on comparing and assessing their performance using a unified framework, and thus few suggestions and guidance for choosing an appropriate method can be provided to the practitioners. Therefore, the performance assessment study for PM-FD methods has become an area of interest in both academia and industry.

The first objective of this thesis is to assess the performance of basic FD statistics. The commonly used two statistics, namely, T^2 and Q are first examined. With the aid of χ^2 distribution, their differences to detect additive and multiplicative faults are revealed and compared under the statistical framework. Due to their low detectability to multiplicative faults, some alternative statistics are investigated.

Based on the basic FD statistics, different PM-FD methods have been proposed to monitor the key performance indicators (KPIs) of static processes, steady-state dynamic processes and dynamic processes including transient states. Thus, the second objective of this thesis is to assess the three classes of KPI-based PM-FD methods. Firstly, existing static methods are sorted into three categories based on the way to partition the KPI-correlated part from the KPI-uncorrelated part. A new EDD index is proposed to assess their performance to detect offsetting, drift and multiplicative faults. Secondly, two dynamic partial least squares (DPLS)-based methods for steady-state dynamic processes are compared, and their performance is assessed using EDD. Furthermore, the KPI-based PM-FD methods for general dynamic processes are introduced, some new developments are given.

Finally, to validate the theoretical developments, a case study on the Tennessee Eastman benchmark process that can be considered as a steady-state dynamic process is performed to assess the two DPLS-based methods. In addition, a real large-scale hot strip rolling mill process is applied to assess the dynamic KPI-based PM-FD methods.

This work was done while I was with the Institute for Automatic Control and Complex Systems (AKS) at the University of Duisburg-Essen, Duisburg, Germany. I would like to express my deepest gratitude to my supervisor, Prof. Dr.-Ing. Steven X. Ding, for all the inspiration and help he provided during the course of the last three and a half years. I am sincerely grateful for his guidance and influence on my scientific research work. I would also like to thank Prof. Peng for his interest in my work. Without his valuable discussions and constructive comments, the thesis cannot have reached the current level.

Furthermore, I would like to express my appreciation to my colleagues, Zhiwen, Dr. Hao, Dr. Shardt, and Prof. Ge for all the impressive discussions and cooperations on my research topic as well as for their patience to go over the draft for this thesis. My special thanks should once again go to Dr. Shardt, who has shared his rich and valuable experiences on academic research and scientific writing.

In addition, I would like to thank Linlin, Changchen, Hao, Minjia, Sihan, Dongmei, Ying, and Yong for their support during my stay in AKS. My thanks also go to all my other AKS colleagues, Tim K., Chris, Shane, Tim D., Sabine, Dr. Köppen-Seliger, Klaus, Ulrich, Dr. Qiu, Dr. Li, and Dr. Jiang as well as my former colleagues, Prof. Lei, Prof. Shen, Prof. Dong, and Prof. Yang for their valuable discussions and helpful suggestions. Without them the completion of this thesis would not have been possible.

Finally, I would like to thank the China Scholar Council (CSC) for funding my stay in Germany.

Kai Zhang

Contents

List of Figures

List of Tables

Abbreviations and notations

Abbreviations

Abbreviation	Expansion
AIC	Akaike Information Criterion
ARMA	Auto-Regressive Moving Average
CCA	Canonical Correlation Analysis
CDF	Cumulative Distribution Function
C-PLS	Concurrent Partial Least Squares
DD	Detection Delay
DDPLS	Direct Dynamic Partial Least Squares
DO	Diagnostic Observer
DPLS	Dynamic Partial Least Squares
EDD	Expected Detection Delay
FAR	False Alarm Rate
FD	Fault Detection
FDR	Fault Detection Rate
FIR	Finite Impulse Response
FMP	Finishing Mill Process
HSMR	Hot Strip Mill Rolling
IDPLS	Indirect Dynamic Partial Least Squares
KPI	Key Performance Indicator
LS	Least Squares
LTI	Linear Time-Invariant
MSPM	Multivariate Statistical Process Monitoring
NIPALS	Nonlinear Iterative PArtial Least Squares
PCA	Principal Component Analysis
PCR	Principal Component Regression
PDF	Probability Distribution Function
PLS	Partial Least Squares
PM	Process Monitoring
PM-FD	Process Monitoring and Fault Detection
PRESS	Predicted REsidual Sum of Squares

PS	Parity Space
SVD	Singular Vector Decomposition
TE	Tennessee Eastman
T-PLS	Total Partial Least Squares
VAR	Vector Auto-Regression

Mathematical notations

Notation	Description
$\forall$	For all
$\sim$	Follow
$\approx$	Approximately equal
$\triangleq$	Defined as
$\gg$	Much larger than
$\to$	From...to
$\mathbb{R}^m$	Set of m-dimensional real vectors
$\mathbb{R}^{m \times n}$	Set of $m \times n$-dimensional real matrices
$\mathbf{I}_m$	m-dimensional identify matrix
$\lVert \cdot \rVert_\mathbf{E}$	Euclidean norm of a vector
$\lVert \cdot \rVert_\mathbf{F}$	Frobenius norm of a matrix
$\lVert \cdot \rVert_2$	2-norm of a matrix
$\lvert \cdot \rvert$	Determinant of a matrix or absolute value
c	A real constant
$\mathbf{y}$	A vector
$\mathbf{y}(i)$	The ith element of $\mathbf{y}$ or the ith sample of $\mathbf{y}$
$\mathbf{y}^i$	i^{th} iteration of $\mathbf{y}$
$\mathbf{Y}$	A matrix
$\hat{\mathbf{y}}$	Estimate of $\mathbf{y}$ or KPI-related part in $\mathbf{y}$
$\tilde{\mathbf{y}}$	Residual of $\mathbf{y}$, or KPI-unrelated part in $\mathbf{y}$
$\mathbf{Y}^T$	Transpose of $\mathbf{Y}$
$\mathbf{Y}^{-1}$	Inverse of a square matrix $\mathbf{Y}$
$\mathbf{Y}^\dagger$	Pseduoinverse of $\mathbf{Y}$
$\mathbf{Y}^\perp$	Orthogonal complement of $\mathbf{Y}$
$\mathbf{f}$	Fault vector
f	Fault magnitude
$\boldsymbol{\Xi}$	Fault direction
λ	Eigenvalue
σ	Singular value or standard derivation

$\mathrm{tr}(\mathbf{Y})$	Trace of $\mathbf{Y}$
$\mathrm{diag}(\mathbf{y})$	A diagonal matrix with non-zeros elements $\mathbf{y}$
$\mathrm{rank}(\mathbf{Y})$	Rank of matrix $\mathbf{Y}$
$\dim\{\cdot\}$	Dimension of a space
$\mathrm{span}\{\mathbf{y}\}$	Space spanned by $\mathbf{y}$
$\oplus$	Direct sum of two vector-spanned spaces
$\otimes$	Kronecker product
$\propto$	Proportional
$\mathrm{E}(\cdot)$	Mean value/vector
$\mathrm{Var}(\cdot)$	Variance value/vector
$\mathrm{Cov}(\cdot)$	Covariance matrix
$\mathrm{prob}(x)$	Probability of x
$\mathcal{N}_m(\mu, \Sigma)$	m-dimensioned Normal/Gaussian distribution with mean μ and covariance matrix Σ
J	Test statistic
J_{th}	Threshold
χ^2_m	Chi-squared distribution with m degrees of freedom
$\chi^2_m(\delta)$	Noncentral χ^2 distribution with m degrees of freedom and noncentrality parameter δ
$\mathcal{F}(a, b)$	$\mathcal{F}$ distribution with a and b degrees of freedom
α	Significance level
$\chi^2_{m,\alpha}$	Confidence value corresponding to α
$\mathcal{F}_\alpha(a, b)$	Confidence value corresponding to α
$\mathcal{W}_m(\Sigma, n)$	Wishart distribution with n degrees of freedom based on m-dimensional covariance matrix Σ
e, $\exp(\cdot)$	Base of natural logarithm, natural exponential function

1 Introduction

1.1 Background and basic concepts

Consider a typical industrial process as shown in Figure 1.1. Control signals sent from the controller are feeded into actuators, where the process input signals are generated. The process is driven by the input signals to achieve the desired output behavior. Finally, sensors convert the output variables as measurement variables, which provide essential information for implementing closed-loop control. It is common for a real process that all these components are subject to disturbances in a stochastic manner. As a result, the input and output signals as well as the measurements are corrupted with noise. An example to this problem is the white noise in measurements, which is due to the accuracy of the sensor and noisy ambient. In reality, such processes are threaten by various faults that may occur in all components. They can not only break the control loop at the process level, but also cause unexpected changes in the plant level. To achieve optimal process operation, these faults should be readily and accurately detected. This, thus, motivates and drives the development of fault detection (FD) methods in both theory and practice. Conceptually, these methods deal with the following task [1–5]:

Fault detection: detection of the abnormal events in the functional units of the process, which can lead to undesired or unacceptable behavior of the whole plant.

It is noted that FD methods are commonly performed at the process level, which means there should be sufficient process knowledge including at least process input and output information. As well known, large-scale processes are ubiquitous features of many chemical, steelmaking and papermaking plants. Such large-scale processes consist of great number of interacting subprocesses which increase the overall control complexity.

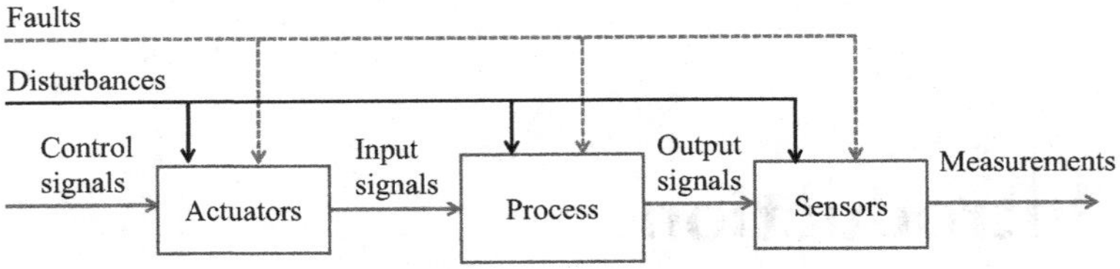

Figure 1.1: Schematic description of an industrial process

Due to increasing demands for quality, a greater emphasis on improving operating performance of these large-scale processes can be observed. This results in strong needs to monitor the process operation at the plant level. Consequently, process monitoring (PM) methods have been extensively reported in the last two decades and widely applied in various industrial plant, such as chemical industry, semiconductor manufacture, steel industry *etc.* A technical description of process monitoring, as given in [14, 18, 77, 85, 111] is

Process monitoring: often referred as statistical process monitoring, generally defined as the use of statistical methods to monitor the operation of the process to improve process quality and productivity

Aiming at PM, two groups of methods are generally used. The first group check the entire process measurements for the purpose of monitoring the performance of the whole plant. Another group pays the attention to the performance of the most important variables. These variables are not always easily measured but can directly indicate the plant operating performance, which has recently been adopted as key performance indicators (KPIs) to analyse the process performance [6, 8]. Hao *et al.* [7], showed that industrial KPIs can be classified into three groups:

- engineering KPIs that refer to the technical performance of the plant, for example, product quality;
- maintenance KPIs that refer to the operating rate and hence maintenance time and costs;
- economic KPIs that refer to business profit, for example, the overall energy consumption or the productivity of a plant.

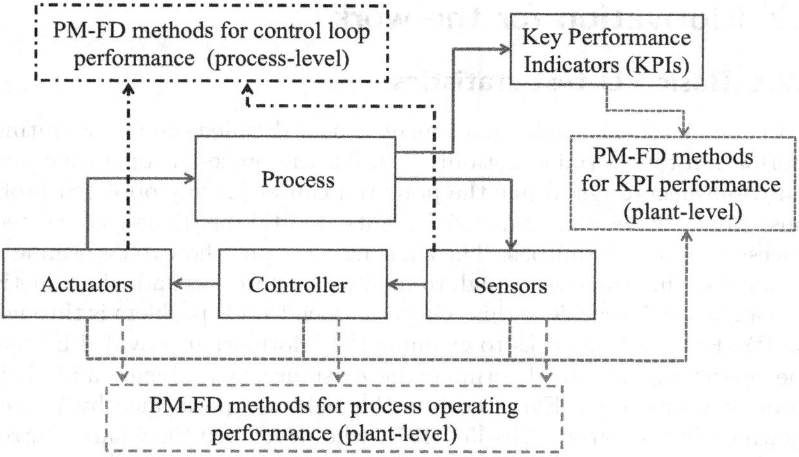

Figure 1.2: Schematic description of PM-FD methods

It has been shown that KPIs are closely related to the measurable process variables, but difficult to be directly measured [8, 28], for example, the concentration in a chemical process or the thickness of a steel roll between two stands in the steel mill process. KPI-based PM methods are primarily developed by applying the online readily measurable variables to track the behavior of KPIs. This kind of approaches have been shown being powerful and effective in detecting process faults that negatively influence KPIs and so enhancing the product quality. It can likewise be seen that KPI-based PM methods are performed at the plant level. Note that although FD and PM methods occur in different levels, from the statistical perspective, there are mixture use of them in literature [40]. It is common that reporting the process as normal or not can also be regarded as determining wether a fault occurred or not in the FD method. In this thesis, in order to avoid the terminological misleading, process monitoring and fault detection (PM-FD) will be adopted to account for plant-level methods. The overall PM-FD issues addressed in industrial plants are structured in Figure 1.2 [13]. Due to the increase in demanding high quality products and high-efficiency performance, this thesis focuses on the KPI-based PM-FD methods.

1.2 Motivation for the work

1.2.1 Basic FD test statistics

Process maintenance and management require detailed process operating information to determine not only whether the process is operating normally, but also to determine the potential causes for any observed problems [118]. In modern industrial plants, multidimensional, correlated process data are ubiquitous. The challenging issue is how to determine if the data are informative enough to monitor the process and which methods can be used to achieve this. One approach to this problem is through the PM-FD [36] that seeks to examine the information provided by routine operating data to determine the existence of problems and their probable root causes. Early work in this field was performed by Walter Shewhart in the early 1920s [53, 107], who developed Shewhart control charts that allows easily tracking of the reliability of telephony transmission systems. Afterwards, this approach has been widely adopted in other technically and physical processes, where a normal distribution is typically assumed. Shewhart charts are easy to create, but are limited to univariate monitoring which does not take into consideration any dependencies between the monitored variables [53]. Driven by the demands of safety and regulation in industrial plants, countless KPI-based PM-FD approaches have been developed for easy tracking of the KPI variable [13]. Due to the stochastic disturbances, as shown in Figure 1.1, using solely the mean of process variables as a sufficient descriptor is dubious. In fact, it would be better to consider the probability distribution of the process variable. The most common solution to this issue is using multivariate statistical techniques, where process variables are assumed to follow multivariate normally distribution. In this framework, multivariate detection statistics are then developed which can simultaneously monitor an ensemble of variables to determine whether the process is behaving properly. For example, a process with two Gaussian distributed process variable is shown in Figure 1.3. A multivariate statistics-based approach seeks to convert the two variables to be an indicator variable that can follow a specific distribution (*e.g.*, χ^2-distribution in Figure 1.3) [130], so that tracking the behavior of the indicator variable would be equivalent to tracking the original multiple variables. Such methods, on the one hand, can avoid separately monitoring the two variables. On the other hand, the dependency between them is taken into account which

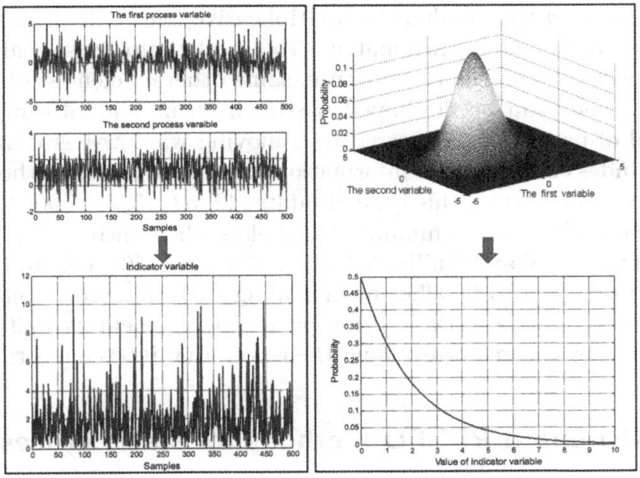

Figure 1.3: Basics of statistical fault detection methods

can improve the FD performance. The transformations/conversions that always refer to the fault detection statistics (J) serve as the core of statistical PM-FD methods. Using some specific probability distributions, a upper threshold J_{th} or two thresholds: the upper one, $J_{th,1}$ and lower one, $J_{th,2}$, are determined. A faulty or normal operating status can then be determined by comparing J with J_{th}. The most widely used detection statistics are T^2- and Q-statistics [2, 48, 52, 54, 69, 73].

In PM-FD field, two types of faults are commonly considered: additive faults, which impact the mean of the variable, and multiplicative faults, which lead to variation in the variance and covariance of the variables [42]. Although additive faults are most commonly assumed in the literature [9, 64], multiplicative faults can also degrade the process efficiency, and impact the safety of the overall system. In previous research, the suitability of T^2 and Q-statistics for detecting these two types of faults was often checked by approximating the fault detection rate (FDR) index using a numerical approximation-based method [71]. However, a theoretical approach to the problem is more required. To establish a clear mathematical foundation for them can lead to their developments and support the implementations in PM-FD methods.

Unlike mean change faults, the multiplicative fault will cause changes in elements of the covariance matrix. To detect the process change that could impact the covariance structure, some other efficient statistics are available. They can be developed based on an individual sample or a sequential of process data covered by a moving window-based approach which includes enough faulty information. Although many methods have been proposed to detect this type of faults [42, 63–65, 67, 100, 102, 103], and some useful tools in communication field such as entropy [107], mutual information [108] and Kullback-Leibler divergence [67, 68, 70, 103, 132] have been reported to be efficient in dealing with this type of change in signals, there is little work focusing on reviewing them as well as comparing them by means of revealing their potential interconnections.

1.2.2 KPI-based PM-FD methods for static process

In static processes, it is assumed that process variables have no autocorrelations, and current KPI measurements can only be influenced by current process measurements. At the same time as the development of fault detection statistics had occurred, work in chemometrics led to the development of new data analysis methods, for example, principal component analysis (PCA) and partial least squares (PLS) [78, 79, 92, 121], which led to increased process efficiencies [25, 30, 35, 36, 122, 125, 127] and understanding [36, 39, 50, 80, 123]. Finally, in the early 1990s, the PLS and PCA methods were combined with T^2- and Q-statistics leading to the development of a new field of PM-FD approaches for static processes [19–21]. The pioneer work was started by MacGregor [15–17], and successively developed by the work of Qin *et al.* [13, 14], and Venkatasubramanian *et al.* [115–117]. These methods are primarily called multivariate statistics process monitoring (MSPM)-based or data-driven methods [13], and can be well structured in the process control framework as shown in Figure 1.2 as plant-level methods. It is shown that they take all the information about the process components (actuators, sensors, controllers, and KPI) in a process control loop into consideration. Thus, they can address different types of process faults. The general procedure is to develop analytical models of normal and faulty operating conditions, onto which the current process data can be projected to give a measure of current process performance [118]. The key difference between the PCA- and PLS-based methods is the way of using the available data space. As shown in Figure 1.2, PCA-based methods monitor the complete data space [11, 14], while PLS-based methods monitor solely a subspace of the complete data space, commonly referred to as the KPI-correlated subspace [2]. Due to the lack of first principles models, MSPM has been quickly adopted by chemical engineers [24, 25, 29, 37]. As well, such methods have been applied to such areas as semiconductor, polymers, iron, and steel processes [10, 26]. Although many different approaches to PCA and PLS have been reported in the literature, few of them follow a unified framework that explicitly utilizes the T^2- and Q-statistics [27, 40, 71].

Over the past few years, great effort has been made on the modification of PLS aiming at improving the KPI-based PM-FD performance. Representative approaches are total PLS (T-PLS) [37] and concurrent PLS (C-PLS) [24]. Despite showing strong applicability in MSPM area, PLS was originally proposed as an alterative of least squares (LS) in linear regression field [38, 39]. The typical linear regression-based methods are studied by Ding *et al.* [6] and Yin *et al.* [40]. Note that a simple method directly decomposing the cross-covariance between process and KPI variables can also solve this problem, while it has not drawn much attention. Finally, it is noted that even though these methods are reported to be practical in industrial application, few of them have been theoretically assessed to determine their performance [27, 40, 41].

In many industrial applications, MSPM methods are used to detect faults, of which the most common application is to detect additive faults, that is, those which change the mean value of the process. The application and assessment of these methods to detect multiplicative faults, which impact the variance or covariance parameters of process variables are rarely considered. In [9], Hao *et al.* have shown, by comparing the original and current formulae for the T^2-statistic, that MSPM methods could be applied to multiplicative faults. However, greater details, specially from a statistical viewpoint, are required before such methods can be applied to detect multiplicative faults. In addition to this approach, many other methods have been proposed for detecting multiplicative faults [64, 66, 67]. Although many improvements on above-mentioned methods in the literature have been reported [24, 30, 37, 80], these methods cannot well address cases that KPI variables are dynamically related to process variables.

1.2.3 KPI-based PM-FD methods for steady-state dynamic processes

For dynamic processes operating in steady state [82], to address the dynamic issue, dynamic PLS (DPLS) models were proposed [12, 83, 84, 88–90, 126], and quickly adopted both in control engineering and PM-FD fields [34, 83, 87, 88, 91, 94]. While PLS models developed using data independent of time, DPLS models are built based on data at current and past time, and attempt to interpret the current KPI using sufficient past process information [12]. Although different DPLS methods were proposed, those that use the augmented process data to model KPI have major focuses [12, 86, 91]. They follow the similar procedure with PLS models, which allows an easy understanding and implementation. The core idea is to extract the useful information from the current and past process data, and combine them to predict the current KPI. Based on how they determine the weighting vector to extract the KPI-relevant information, two DPLS methods are obtained, the direct DPLS (DDPLS) method, which uses different weighting vectors [88], and the indirect D-PLS (IDPLS) method, which uses the constant weighting vector [12]. Although the two DPLS methods are extensively applied, to date there has been no detailed study on them in terms of computation, convergence characteristic, and potential relationships. Furthermore, since the development of the nonlinear iterative partial least squares (NIPALS) method for PLS models [21, 89, 92], extending it to the above two DPLS methods would be useful, because it would avoid an eigenvalue decomposition performed on a high dimensional matrix. As well, it would make it straightforward to identify and understand the difference between the two DPLS methods.

Application of DPLS to KPI-relevant PM-FD was motivated by the successful application of PLS-based methods [88], where it assumes that the scores of DPLS that represent the KPI-relevant information in process data are time independent. However, this is not always the case in actual circumstance. Recently, Li *et al.* proposed an approach that fits a vector autoregression (VAR) model to the resulting scores [12]. The VAR model is then adopted to obtain the residual vector for KPI-based PM-FD. This method was shown to be effective and extended to dynamic PCA-based methods [93]. To assess the performance of DPLS methods for PM-FD, this approach will be incorporated into DPLS methods.

1.2.4 KPI-based PM-FD methods for dynamic process

In the dynamic case, irrespective of whether the process is at steady state or transient, the relationship between KPIs and process variables can be represented using a state-space model that well describes the process dynamics [33, 97, 98]. By means of the developed state-space model, a residual generator using either parity space (PS) or a diagnostic observer (DO) can be developed to implement PM. Using a data-driven approach, Ding *et al.* [32] have proposed the data-driven PS and DO, which only require the process data. The vectors used for generating the residuals are called the kernel representation of a process [2]. Applications of this method include the data-driven PM of a complex hot steel mill rolling (HSMR) process [6], wherein all the process variables were employed as inputs and all KPIs as outputs to develop a numerically stable solution to the problem, and quality-related FD for a paper mill process [119]. Recently, Shardt *et al.* [8] extended this approach by incorporating a soft-sensor-based PM to deal with infrequent KPI data. Although these approaches are simple to understand, they have not taken into consideration the dynamics in each individual subprocess.

1.2.5 Performance evaluation

Not only the theoretical comparison, but the performance evaluation also is important. For one thing, the evaluation results can benefit the plant engineers to select a method. For another, it may motivate the development of some more efficient methods. This motivates to address a further issue in this thesis, namely performance evaluation of the those methods under consideration. Concretely, it deals with (a) defining evaluating indices in the statistical framework; (b) evaluating the detectability of detection statistics obtained by the methods using these indices [71]. Evaluation of the methods under a universal set of benchmark has not yet drawn sufficient attention. The most common and frequently used evaluation scheme for PM-FD approaches is to check two types of alarms caused when the monitoring index has crossed its corresponding threshold [42–44]. A false alarm is an alarm raised without the presence of any abnormality within the process, while a missed alarm is a signal which is not triggered in the presence of fault. They are also widely termed as type I and type II errors in the industrial sphere [3, 43, 44]. In assessing the performance of FD methods, three different aspects need to be considered: the false alarm rate (FAR), which examines the performance of the method in normal operating conditions; the FDR, which considers the performance during faulty conditions; and the detection delay (DD), which measures the time delay before a fault is detected. In most cases, only the first two metrics are considered [10, 29], while using all three is rarer [22, 40]. These methods are often defined either using a probabilistic approach [2, 43–45] or using a numerical approximation approach [40]. Ding has defined them based on a statistical framework [2], while Yin *et al.* defined them based on how to practically and easily compute the definitions [40, 46]. Aiming at alarm management, Yang *et al.* investigated the analytical probability distribution of fault detection delay using the classic general likelihood ratio and cumulative sum-based fault detection schemes [47]. However, it does not seem useful for methods like those applied in this thesis, *i.e.* T^2- and Q-statistics, owing to their high complexities. Furthermore, since the FAR does not equal zero, the traditional definition that uses the first alarm time instant as the detection moment and calculate the detection delay seems also unreasonable. Thus, a new statistical tool for measuring the delay of a detection should be defined, so that, it, on the one hand, can tell whether the method can detect the fault or not, and on the other hand, can estimate the delayed time for an effective detection [71].

1.3 Objectives

With the motivation to deal with the theoretical problems and meet the practical needs, the objectives of this thesis are:

- to present a fundamental study on the commonly used FD statistics (*e.g.*, T^2- and Q-statistics) including their statistical properties, performance for dealing with different types of faults.
- to group and compare the KPI-based MSPM methods for static processes;
- to review and compare the two DPLS-based PM-FD methods for steady-state dynamic processes,
- to develop, for the dynamic case, a dynamic method that considers the relationships between the individual subsprocesses, as well as between the KPIs and process variables.

which are driven by the theoretical aspect. As well, the thesis seeks

- to define a new index to assess MSPM methods when detecting constant additive, drift and multiplicative faults;
- to assess the performance of DPLS methods when applied to KPI-based PM-FD, and
- to apply the proposed approaches to the benchmark process and a real HSMR process.

which are the practical goals of the thesis.

1.4 Outline of the thesis

The thesis structures the chapters as shown in Figure 1.4. The key contributions of each chapter are briefly summarized as follows.

Chapter 2: Basics of fault detection and performance assessment techniques

This chapter presents the problem formulation of PM-FD in industrial processes, including two types of faults, *i.e.* additive and multiplicative faults, and how they affect process measurements. In addition, the widely adopted performance evaluation indices, such as FAR, FDR, and DD, are reviewed. Finally, a new index called expected detection delay (EDD) is proposed to give more accurate evaluation results for the detection delay given by a method for statistical processes. Its applicability to additive and drift faults will be proven and demonstrated using numerical simulations.

Chapter 3: Common test statistics for fault detection

This chapter examines these two statistics in light of the FDR index to determine their application to additive and independent multiplicative faults. Their different impact on computing FDR is shown. As well, their drawbacks to detect multiplicative faults are investigated. It is followed by a review and comparison of the other existing statistics developed for detecting multiplicative. A numerical simulation is used to show the theoretical results.

Chapter 4: KPI-based PM-FD methods for static processes

In this chapter, the KPI-based multivariate statistical PM-FD methods for linear static processes are surveyed and evaluated in the multivariate statistics framework. Based on their computational characteristics, the possible methods are broadly classified into three categories: direct, linear regression-based, and PLS-based. The comparison study in aspects of their interconnections, geometric properties, and computational costs are shown, and finally their performance for PM-FD of KPIs is assessed in terms of EDD. A numerical simulation will be used to demonstrate the evaluation results.

Chapter 5: KPI-based PM-FD methods for steady-state dynamic processes

The two most interesting DPLS methods that are distinguished based on how they design the weighting vectors are compared in this chapter. Furthermore, in order to improve the performance of these methods, a new NIPALS approach is proposed to avoid the eigenvalue decomposition in the original matrices. Finally, in order to test the application of these two methods to KPI-based PM-FD, the EDD index is developed for these methods and tested on a numerical case study.

Chapter 6: KPI-based PM-FD methods for dynamic processes

In this chapter, the methods using state space to represent the process dynamics are introduced. Firstly, the DO-based method is given. Then for processes with explicit multiple control loops, a subprocess-based method will be presented that takes into account the dynamics in each subprocess to build the PM-FD model for KPI.

Chapter 7: Benchmark study and industrial application

In this chapter, the methods presented in Chapters 5 and 6 are illustrated using the Tennessee Eastman (TE) and HSMR processes.

Chapter 8: Conclusions and future work

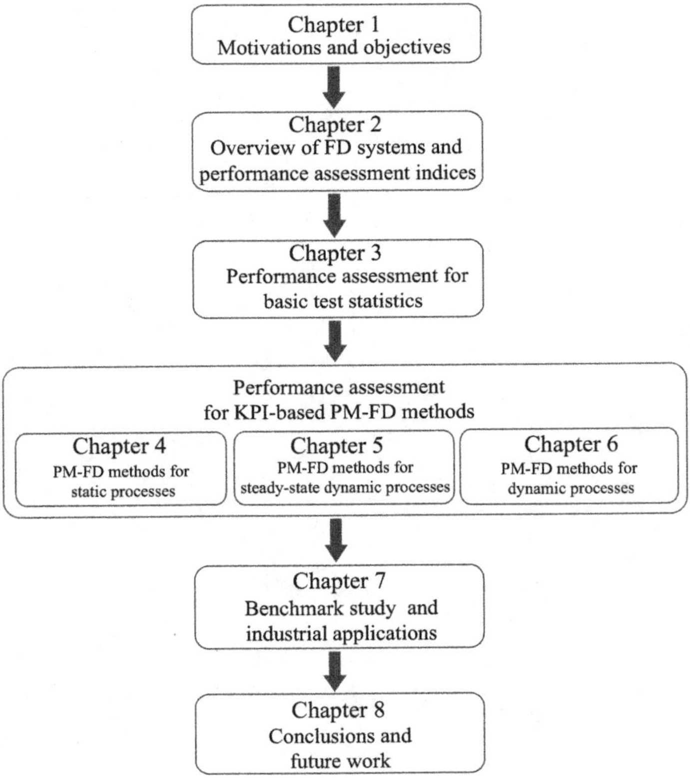

Figure 1.4: Structure of the thesis

2 Basics of fault detection and performance evaluation techniques

Modern large-scale plants are composed of several interconnected process units. Each unit can be approximately modelled using an linear time invariant (LTI) system. Such systems are commonly classified into static and dynamic classes. Mathematical descriptions of them are different, meanwhile, the occurrence of additive and multiplicative faults will impact them in different ways. As the fundamental issues of PM-FD, these two problems will be addressed in the first part of this chapter. Performance evaluation for PM-FD methods is primarily conducted using some standard indices. In the second part, commonly used indices and a new index will be presented.

2.1 Technical description of static processes

In a control system, one of the most important blocks is the sensor, which measures the process value for the control purpose. Due to uncertain environment around the process, the process value returned by the sensor will be affected by various disturbances that will cause the value to deviate from the true value. Furthermore, there may be faults in the process which will disturb the measurement value. For static processes, the observed process value returned by the sensor can be split into three parts: the mean value, changes to the process, and random fluctuations. Therefore, consider a number of sensors, and let the sensor variables be a vector denoted by $\mathbf{y}_{obs} \in \mathbb{R}^m$ and written as [67]

$$\mathbf{y}_{obs} = \boldsymbol{\mu}_y + \underbrace{\mathbf{Az} + \boldsymbol{\nu}}_{\mathbf{y}} \tag{2.1}$$

where μ_y is the mean (or expected) process value, $\mathbf{A} \in \mathbb{R}^{m \times m}$ denotes the process parameter matrix, $\mathbf{z} \in \mathbb{R}^m \sim \mathcal{N}_m(0, \mathbf{I}_m)$, $\mathbf{Az}$ represents the relevant process changes, $\boldsymbol{\nu} \in \mathbb{R}^m \sim \mathcal{N}_m(0, \Sigma_\nu)$ denotes the measurement noise, Σ_ν is a diagonal matrix. For such a system, the covariance structure is written as $\Sigma_y = \mathbf{A}\mathbf{A}^T + \Sigma_\nu$ with $\mathbf{y} \sim \mathcal{N}_m(0, \Sigma_y)$.

In general, process faults are classified as either additive or multiplicative faults. An additive fault only changes the mean value of the process and can be modelled as:

$$\mathbf{y}_{obs,f} = \mu_{y,f} + \mathbf{Az} + \boldsymbol{\nu} = \mu_y + \underbrace{\boldsymbol{\Xi} f + \mathbf{Az} + \boldsymbol{\nu}}_{\mathbf{y}_f} \tag{2.2}$$

where $\mathbf{y}_f \sim \mathcal{N}_m(\boldsymbol{\Xi} f, \Sigma_y)$, $\boldsymbol{\Xi} \in \mathbb{R}^m$ is a vector of unit length denoting the fault direction, and $f \geq 0$ denotes the fault magnitude. Additive faults represent changes in a sensor's accuracy. On the other hand, multiplicative faults result from changes in the covariance structure. A typical case is the parameter change occurring in $\mathbf{A}$. For such faults, the measurements are represented as

$$\mathbf{y}_{obs,f} = \mu_y + \underbrace{(\mathbf{A} + \Delta\mathbf{A})\,\mathbf{z} + \boldsymbol{\nu}}_{\mathbf{y}_f} \tag{2.3}$$

where $\mathbf{y}_f \sim \mathcal{N}_m(0, \Sigma_{y,f})$ with $\Sigma_{y,f} = (\mathbf{A} + \Delta\mathbf{A})(\mathbf{A} + \Delta\mathbf{A})^T + \Sigma_\nu$. $\Sigma_{y,f}$ can be simply converted to $\Sigma_{y,f} = \mathbf{M}\Sigma_y\mathbf{M}$ [2, 9, 112, 114], with $\mathbf{M} \in \mathbb{R}^{m \times m}$ and $M_{i,j}$ representing the change of variance and covariance. They are mainly referred to abnormal changes like the degradation of working components in processes. Another common type of multiplicative fault is an independent multiplicative fault that causes changes in the diagonal terms of the covariance matrix, Σ_ν. Such a change represents a change in the precision of the sensor. This fault can be modeled as:

$$\mathbf{y}_{obs,f} = \mu_y + \underbrace{\mathbf{Az} + \boldsymbol{\nu}_f}_{\mathbf{y}_f} \tag{2.4}$$

where $\mathbf{y}_f \sim \mathcal{N}_m(0, \Sigma_f)$, with $\Sigma_{y,f} = \mathbf{A}\mathbf{A}^T + \Sigma_{\nu f}$. Furthermore, $\Sigma_{y,f}$ is likewise represented as $\Sigma_{y,f} = \mathbf{M}\Sigma_y\mathbf{M}$, where $\mathbf{M} = \mathrm{diag}(M_1, \cdots, M_m)$, M_i denotes the variance change in the ith variable.

2.2 Technical description of dynamic processes

LTI systems are widely used to describe a dynamic process using a state-space model. The nominal form of state-space representation of a discrete-time LTI system is

$$\begin{aligned} \mathbf{x}(k+1) &= \mathbf{A}\mathbf{x}(k) + \mathbf{B}\mathbf{u}(k), \mathbf{x}(0) = \mathbf{x}_0, \\ \mathbf{y}(k) &= \mathbf{C}\mathbf{x}(k) + \mathbf{D}\mathbf{u}(k) \end{aligned} \tag{2.5}$$

where $\mathbf{x} \in \mathbb{R}^n$ is the state vector, $\mathbf{x}_0$ is the initial condition of the system, $\mathbf{u} \in \mathbb{R}^l$ and $\mathbf{y} \in \mathbb{R}^m$ are the input and output vectors. $\mathbf{A} \in \mathbb{R}^{n\times n}$ is the state transition matrix, $\mathbf{B} \in \mathbb{R}^{n\times l}$ is the input matrix, $\mathbf{C} \in \mathbb{R}^{m\times n}$ is the output matrix, and $\mathbf{D} \in \mathbb{R}^{m\times l}$ is the feed-through matrix. Considering the stochastic disturbance, the system is written as

$$\begin{aligned} \mathbf{x}(k+1) &= \mathbf{A}\mathbf{x}(k) + \mathbf{B}\mathbf{u}(k) + \boldsymbol{\eta}(k) \\ \mathbf{y}(k) &= \mathbf{C}\mathbf{x}(k) + \mathbf{D}\mathbf{u}(k) + \boldsymbol{\nu}(k) \end{aligned} \tag{2.6}$$

where $\boldsymbol{\eta}(k)$ and $\boldsymbol{\nu}(k)$ are Gaussian process, independent of $\mathbf{x}_0$ and $\mathbf{u}(k)$ with

$$\mathrm{E}\begin{bmatrix} \boldsymbol{\eta}(i)\boldsymbol{\eta}^T(j) & \boldsymbol{\eta}(i)\boldsymbol{\nu}^T(j) \\ \boldsymbol{\nu}(i)\boldsymbol{\eta}^T(j) & \boldsymbol{\nu}(i)\boldsymbol{\nu}^T(j) \end{bmatrix} = \begin{bmatrix} \Sigma_\eta & S_{\eta\nu} \\ S_{\nu\eta} & \Sigma_\nu \end{bmatrix}\delta_{i,j}, \delta_{i,j} = \begin{cases} 1, i = j \\ 0, i \neq j \end{cases}$$
$$\mathrm{E}(\boldsymbol{\eta}(i)) = 0, \mathrm{E}(\boldsymbol{\nu}(i)) = 0.$$

The faults occurring in such kind of systems can be modeled in various ways. A commonly used one is to extend Eq. (2.6) to the form of

$$\begin{aligned} \mathbf{x}(k+1) &= \mathbf{A}\mathbf{x}(k) + \mathbf{B}\mathbf{u}(k) + \mathbf{E}_f\mathbf{f}(k) + \boldsymbol{\eta}(k) \\ \mathbf{y}(k) &= \mathbf{C}\mathbf{x}(k) + \mathbf{D}\mathbf{u}(k) + \mathbf{F}_f\mathbf{f}(k) + \boldsymbol{\nu}(k) \end{aligned} \tag{2.7}$$

where $\mathbf{f}(k) \in \mathbb{R}^{d_f}$ is a unknown vector that denotes all possible faults and will be zero in the fault-free case, $\mathbf{E}_f$ and $\mathbf{F}_f$ are properly dimensioned indicting (1) where a fault occurs; (2) how it impact the system dynamics. According to the location, the faults are divided into three classes [1, 112, 114]:

- Sensor faults, $\mathbf{f}_S$: faults that directly impact the process measurements;
  e.g., let $\mathbf{E}_f = \mathbf{0}$ and $\mathbf{F}_f = \mathbf{I}_l$, $\mathbf{y}_f = \mathbf{y}^* + \mathbf{f}_S$.

- Actuator faults, $\mathbf{f}_A$: faults that could cause changes in actuators; *e.g.*, let $\mathbf{E}_f = \mathbf{B}$ and $\mathbf{F}_f = \mathbf{D}$, $\mathbf{u}_f = \mathbf{u}^* + \mathbf{f}_A$.
- Process faults, $\mathbf{f}_P$: faults that indicate malfunctions within the processes.
  e.g., let $\mathbf{E}_f = \mathbf{E}_p$, $\mathbf{F}_f = \mathbf{F}_p$, $\mathbf{x}_f = \mathbf{x} + \mathbf{f}_P$.

Depending the way how they affect the system dynamics, the faults introduced above are additive faults. They will affect the first-order statistics of output vector, *i.e.* the mean vector of $\mathbf{y}$. It is worth noting that additive faults will not affect the system stability. In practice, there exists another type of fault that is modelled as the change in parameter matrices of Eq. (2.6)

$$\begin{aligned} \mathbf{x}(k+1) &= (\mathbf{A} + \Delta\mathbf{A})\,\mathbf{x}(k) + (\mathbf{B} + \Delta\mathbf{B})\,\mathbf{u}(k) + \boldsymbol{\eta}(k) \\ \mathbf{y}_f(k) &= (\mathbf{C} + \Delta\mathbf{C})\,\mathbf{x}(k) + (\mathbf{D} + \Delta\mathbf{D})\,\mathbf{u}(k) + \boldsymbol{\nu}(k) \end{aligned} \tag{2.8}$$

where $\Delta\mathbf{A}$, $\Delta\mathbf{B}$, $\Delta\mathbf{C}$ and $\Delta\mathbf{D}$ represent the multiplicative fault in system parameters. Compared with the additive fault, this type of fault can influence the second-order statistics of output data. Figure 2.1 shows the two types of faults using a two-variable case, where it can be observed that the additive fault (Eqs. (2.2) and (2.7)) only affects the mean of the data while the multiplicative fault (Eqs. (2.3), (2.4) and (2.8)) does not influence the mean of the measured data, but solely impact the variance.

2.3 FD performance evaluation indices

2.3.1 FDR and FAR

Designing FD methods for monitoring $\mathbf{y}$ consists of (1) defining the detection (test) statistics J with its corresponding threshold J_{th} and (2) comparing the online realization of J, *i.e.* $J(\mathbf{y}(k))$ against J_{th} to make the decision: faulty or fault-free. For example, the extensively used T^2 test statistic is designed as $J_{T^2} = \mathbf{y}^T \Sigma_y^{-1} \mathbf{y}$. The associated J_{th} is commonly of the form: $J_{th,T^2} = \chi^2_{m,\alpha}$. A comprehensive study on J_{T^2} and J_{th,T^2} as well as other statistics will be provided in Chapter 3.

Figure 2.2 shows a typical data display, by which the essential fault detection performance are schematically illustrated. A false alarm occurs when an alarm is announced under normal operating condition. Fault

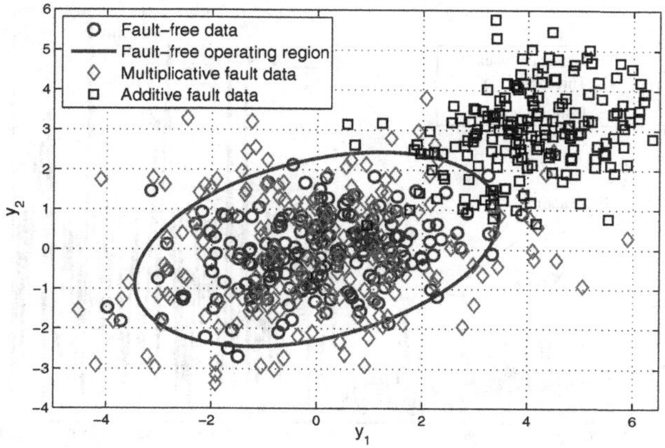

Figure 2.1: Demonstration of additive and multiplicative faults

detection alarm represents an effective alarm issued while there exists a fault [1]. From the probability point of view, FAR and FDR can be correspondingly defined, as they stand for the occurrence probabilities of false alarms and successful fault detection [1]. For the constant fault case, the two definitions are

$$\begin{aligned} \text{FAR} &= \text{prob}\left(J > J_{th} | f = 0\right) \\ \text{FDR} &= \text{prob}\left(J > J_{th} | f = c\,(\neq 0)\right) \end{aligned} \tag{2.9}$$

Yin *et al.* have given the following estimates [40]:

$$\begin{aligned} \text{FAR} &= \frac{\text{Number of samples } (J > J_{th} \,|\text{fault} - \text{free})}{\text{total fault} - \text{free samples}} \\ \text{FDR} &= \frac{\text{Number of samples } (J > J_{th} \,|\text{faulty})}{\text{total faulty samples}} \end{aligned} \tag{2.10}$$

Although widely applied in practice, this method fails when the fault is time varying, for example, in the case of a drift fault, which causes changes in $\mathbf{y}$ slowly. Two FDR-like indices were proposed for quantifying

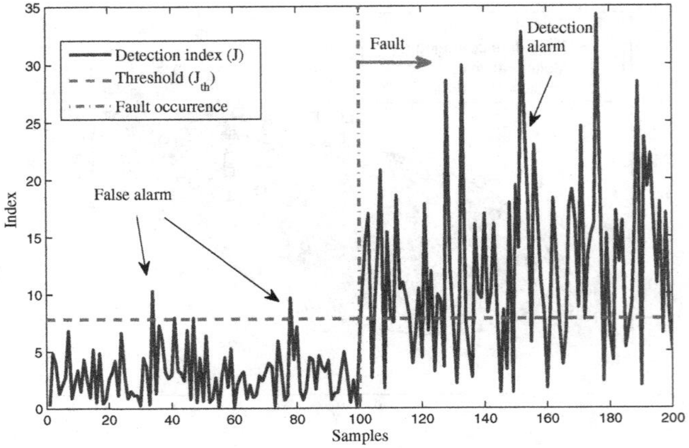

Figure 2.2: Demonstration of false alarm rate and fault detection rate

the probability that $\mathbf{y}(k)$ is faulty [120]:

$$\begin{aligned} \text{prob}\left(\text{fault}\left|\mathbf{y}\left(k\right)\right.\right) &= \text{prob}\left(J\left(k\right) > J\left(\mathbf{y}^{*}\right)\left|\mathbf{y}^{*} \in \mathbf{Y}_{tr}\right.\right) && \text{(2.11a)} \\ \text{prob}\left(\text{fault}\left|\mathbf{y}\left(k\right)\right.\right) &= \exp\left(-\frac{J_{th}}{\varsigma J\left(k\right)}\right) && \text{(2.11b)} \end{aligned}$$

where $J\left(k\right)$ is shorthand for $J\left(\mathbf{y}(k)\right)$, $k \geq k_f$ with k_f denoting the fault occurring time instance, $\mathbf{Y}_{tr}$ denotes the normal training dataset and $\varsigma > 0$ is a tuning parameter, $\exp(\cdot)$ denotes the exponential function. It can be observed that Eq. (2.11a) does not depend on J_{th} and, thus, the calculated value cannot be used to report the fault. In Eq. (2.11b) $\text{prob}\left(\text{fault}\left|\mathbf{y}\left(k\right)\right.\right)$ approaches 1 only given a significantly large $\varsigma J(k)$, tuning of ς will be a trade-off to fulfil different demands, which can lead to difficulties in implementing this method. From the theoretical viewpoint, it should be assumed that in the case of a constant fault, the FAR at each time k should be constant, namely, $\text{FDR}(k) = c \; \forall k > k_f$. Considering the stochastic nature of $\mathbf{y}$, the methods in Eq. (2.11) cannot ensure a constant probability value for a constant additive fault, and the situation will be worse for a multiplicative fault that can cause significant changes in the variances or covariance of $\mathbf{y}$. Therefore, they are not appropriate for estimating FDR(k). In this chapter, another method that considers the distribution of J is proposed [71, 128],

$$\mathrm{FDR}\,(k) = \int_{J_{th}}^{\infty} f_{J,k}\,(x)dx = 1 - F_{J,k}\,(J_{th}) \tag{2.12}$$

where $f_{J,k}$ and $F_{J,k}$ denote the probability density function (PDF) and the cumulative distribution function (CDF) of J at time k. This method avoids the weakness mentioned in Eq. (2.11), and can be easily realized for constant additive faults when adopting the T^2- and Q-statistics.

2.3.2 Expected detection delay

Although extensively implemented, FDR can only reflect the detectable probability of the PM-FD index for the fault with fixed parameters, but cannot tell whether the fault could be instantaneously detected or not. In addition, when the fault is successfully detected, the time taken to detect the fault is also important. Therefore, in this part, a new index, called EDD, is proposed to deal with this issue. If DD is defined as a random variable that shows the possible time interval between the occurrence of the fault and the successful detection of it.

Let $\mathcal{J}$ denote DD, $\mathcal{J}$ may take the value j, where $j = 0, 1, 2, ...\infty$. The probability that $\mathcal{J}$ takes the value j is based on the fact that $J\,(t_f) \leq J_{th}, J\,(k_f + 1) \leq J_{th}, ..., J\,(k_f + j) > J_{th}$ and is shown as

$$\mathrm{prob}\,(\mathcal{J} = j) = \begin{cases} \left(\prod\limits_{k=0}^{j-1} (1 - \mathrm{FDR}\,(k_f + k))\right) \mathrm{FDR}\,(k_f + j)\,, \text{for } j \geq 1 \\ \mathrm{FDR}\,(k_f)\,, \text{for } j = 0 \end{cases} \tag{2.13}$$

where k_f is the fault occurrence time as defined above. EDD takes the expectation of $\mathcal{J}$ based on its distributional information.

Consider the special case where the fault is constant. In this case, $\forall k$, FDR(k) = FDR, where FDR is obtained using Eq. (2.12). Theorem 2.1 shows that for a constant additive fault the expected detection delay provides a valid approach.

Theorem 2.1. *For a constant fault with* FDR(k) $= c < 1\ \forall k$, *then* $\sum\limits_{j=1}^{\infty} \mathrm{prob}\,(\mathcal{J} = j) = 1$.

Proof. Based on Eq. (2.13), we can obtain that

$$\begin{aligned}\sum_{\forall j} \mathrm{prob}\,(\mathcal{J} = j) &= \lim_{n\to\infty} \left(\mathrm{FDR} + (1-\mathrm{FDR})\,\mathrm{FDR} + \cdots + (1-\mathrm{FDR})^n\mathrm{FDR}\right) \\ &= \lim_{n\to\infty} \left(\mathrm{FDR}\frac{1-(1-\mathrm{FDR})^n}{1-(1-\mathrm{FDR})}\right) \\ &= 1\end{aligned} \tag{2.14}$$

□

The above conclusion and Theorem 2.1 demonstrate that the new definition can indicate the PDF of DD for a constant fault. In this case, indeed, $\mathcal{J}$ obeys the geometric distribution [54]. Thus, the expectation of DD is obtained by

$$\mathrm{EDD} \triangleq \mathrm{E}\,(\mathrm{DD}) = \sum_{j=0}^{\infty} j\mathrm{prob}\,(\mathcal{J} = j) = \sum_{j=0}^{\infty} j(1-\mathrm{FDR})^j\mathrm{FDR} = \tfrac{1-\mathrm{FDR}}{\mathrm{FDR}} \tag{2.15}$$

From Eq. (2.15), it can be seen that the EDD index indicates the expected time delay given by a method for detecting a fault. As shown in Figure 2.3, if FDR $\to 1$, that leads to EDD $\to 0$ and the detection results are demonstrated in the bottom left figure; else if FDR $\to 0$, EDD approaches infinity, then top left figure of Figure 2.3 shows the detection in this case. Correspondingly, the middle figure gives the intermediate results. Obviously, the result is consistent with the actual situation, and in other words, the new definition of EDD is right for this particular case. This approach allows users to select an appropriate upper limit for the acceptable detection delay. If the computed EDD value is greater than the threshold, then we can say that the method cannot detect the fault. Practically, it is written as $J_{th,\mathrm{EDD}} = (1-\mathrm{FAR})/\mathrm{FAR} \approx (1-\alpha)/\alpha$ with α denoting a pre-specified significance level [71].

As assumed in this work, for constant faults, irrespective of whether the fault be additive or multiplicative, FDR is constant, thus Eq. (2.12) could be directly used. While for the drift fault case, the FDR will be time-varying. Figure 2.4 shows an example using J_{T^2} to detect this type of fault. It can be observed that as $f(k)$ monotonically increases, the calculated FDR(k) likewise increases. Theorem 2.2, which follows, gives the probabilistic property of EDD for this type of fault.

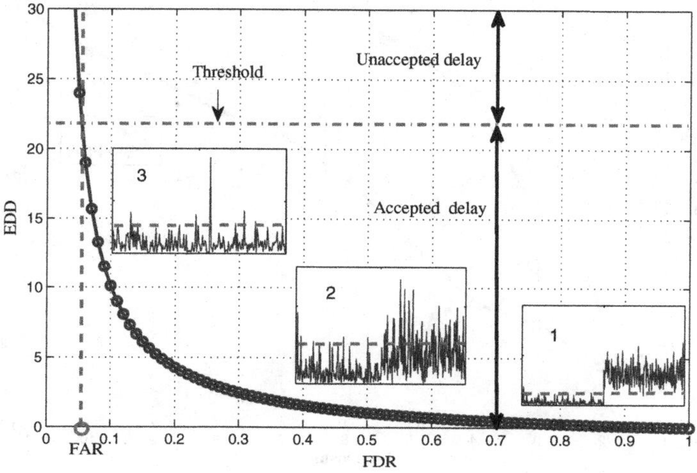

Figure 2.3: Schematic description of detection delay using FAR and FDR

Theorem 2.2. *For a drift fault occurring at time instance k_f, if there exists a time instance k_s that $\forall k \geq k_s$, $\mathrm{FDR}(k) = \mathrm{FDR}(k_s)$, then* $\sum_{j=0}^{\infty} \mathrm{prob}(\mathcal{J} = j) = 1$.

Proof. Let $\mathcal{J} = s + 1$ denote the event that the fault can be detected at the $k_s + 1$ time instance. Since $\mathrm{FDR}(k_s+1) = \mathrm{FDR}(k_s)$, $\mathrm{prob}(\mathcal{J} = s + 1)$ is calculated as

$$\begin{aligned} \mathrm{prob}(\mathcal{J} = s+1) &= \tfrac{\mathrm{prob}(\mathcal{J}=s)}{\mathrm{FDR}(k_s)} (1 - \mathrm{FDR}(k_s))\, \mathrm{FDR}(k_s+1) \\ &= \mathrm{prob}(\mathcal{J} = s)(1 - \mathrm{FDR}(k_s)) \end{aligned} \tag{2.16}$$

This leads to $\mathrm{prob}(\mathcal{J} = s + \tau) = \mathrm{prob}(\mathcal{J} = s)(1 - \mathrm{FDR}(k_s))^{\tau}$. Then,

$$\sum_{j=s}^{\infty} \mathrm{prob}(\mathcal{J} = j) = \mathrm{prob}(\mathcal{J} = s) \sum_{\tau=1}^{\infty} (1 - \mathrm{FDR}(k_s))^{\tau} = \frac{\mathrm{prob}(\mathcal{J} = s)}{\mathrm{FDR}(k_s)} \tag{2.17}$$

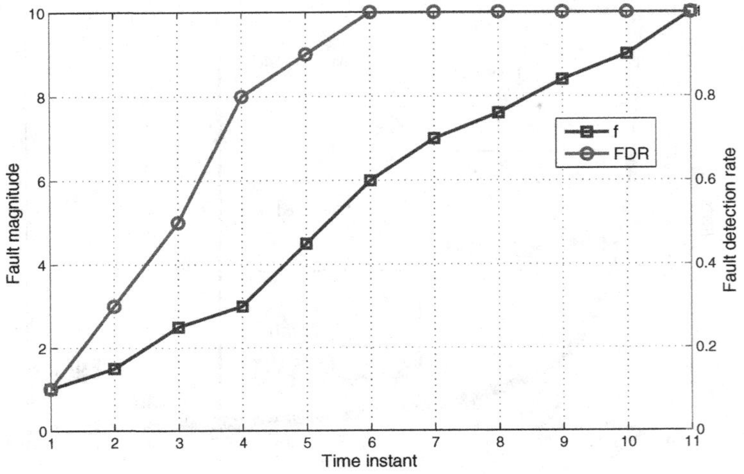

Figure 2.4: An example with FDR for a drift fault

Note that

$$\mathrm{prob}\left(\mathcal{J}=s\right)=\frac{\mathrm{prob}\left(\mathcal{J}=s-1\right)}{\mathrm{FDR}\left(k_s-1\right)}\left(1-\mathrm{FDR}\left(k_s-1\right)\right)\mathrm{FDR}\left(k_s\right) \quad (2.18)$$

Thus,

$$\begin{aligned}\sum_{j=s-1}^{\infty}\mathrm{prob}\left(\mathcal{J}=j\right)&=\sum_{j=s}^{\infty}\mathrm{prob}\left(\mathcal{J}=j\right)+\mathrm{prob}\left(\mathcal{J}=s-1\right)\\&=\frac{\left(\frac{\mathrm{prob}(\mathcal{J}=s-1)}{\mathrm{FDR}(k_s-1)}(1-\mathrm{FDR}(k_s-1))\mathrm{FDR}(k_s)\right)}{\mathrm{FDR}(k_s)}+\mathrm{prob}\left(\mathcal{J}=s-1\right)\\&=\frac{\mathrm{prob}(\mathcal{J}=s-1)}{\mathrm{FDR}(k_s-1)}\end{aligned} \quad (2.19)$$

Using Eqs. (2.17) and (2.19) iteratively gives $\sum_{j=0}^{\infty}\mathrm{prob}\left(\mathcal{J}=j\right)=\frac{\mathrm{prob}(\mathcal{J}=0)}{\mathrm{FDR}(k_f)}=\frac{\mathrm{FDR}(k_f)}{\mathrm{FDR}(k_f)}=1.$ □

Thus, like Theorem 2.1, prob($\mathcal{J}=j$) $\forall j$, can be adopted to describe the PDF of DD for this type of fault. Then, Eq. (2.15) is also valid to calculate the EDD. Note in the case that $\mathrm{FDR}\left(k_s\right)=1$ as show in Figure 2.4, despite $\text{FDR}(k_s+i)<1$ with $i>1$, the experiment should be stopped at time k_s. Using Eq. (2.19) and the fact that $\text{prob}(\mathcal{J}=s)=\frac{\text{prob}(\mathcal{J}=s-1)}{\text{FDR}(k_s-1)}(1-\text{FDR}(k_s-1))$, it is obtained that $\sum_{j=0}^{\tau}\text{prob}(\mathcal{J}=s-j)=\frac{\text{prob}(\mathcal{J}=s-\tau)}{\text{FDR}(k_s-\tau)}$. Let $\tau=s$, $\sum_{j=0}^{s}\text{prob}(\mathcal{J}=j)=\frac{\text{prob}(\mathcal{J}=0)}{\text{FDR}(k_f)}=1$. Therefore, for this type of drift fault, EDD can also be usable.

Remark 2.1. *When calculating EDD, for FD methods that there is more than statistic involved, the FDR value can be determined using the maximum of them.*

2.4 Simulation results

A simple numerical model is used to compare the results given by EDD and the numerical approximation approach for J_{T^2}. Assume that $\mathbf{y}$ is normally distributed with zero mean and covariance matrix

$$\Sigma_y=\begin{bmatrix}2 & 0.5 & 0.3\\ 0.5 & 1 & 0.2\\ 0.3 & 0.2 & 0.5\end{bmatrix} \tag{2.20}$$

Three fault scenarios are examined: a constant additive fault, a drift fault, and a multiplicative fault.

In the first case, a constant additive fault occurs in $\mathbf{y}_2$, where $\Xi=[\,0\ \ 1\ \ 0\,]^T$ and f changes from 1 to 8. For the numerical approximation-based method, the simulation is run 100 times. In each run, the detection delay is recorded. The mean value is finally calculated for comparison. EDD is calculated using Eq. (2.15). Figure 2.5 shows the comparison result. Note that, although the two methods deliver comparable results, the numerical approximation approach-based results contain more fluctuations. In the second case, a drift fault occurs in $\mathbf{y}_3$. $\Xi=[\,0\ \ 0\ \ 1\,]$, and $f(k)=\rho k$ with ρ varying from 0.1 to 0.8. The comparison results are presented in Figure 2.6, where the two methods match well. EDD is calculated in each run when FDR(k) equals 1. In the third case, a multiplicative fault occurs in the system. $\mathbf{M}=\text{diag}(M_1,1,M_3)$, with M_1 and M_3 varying from 1 to 50. The results are shown in Figure 2.7. We can see that, for both test statistics,

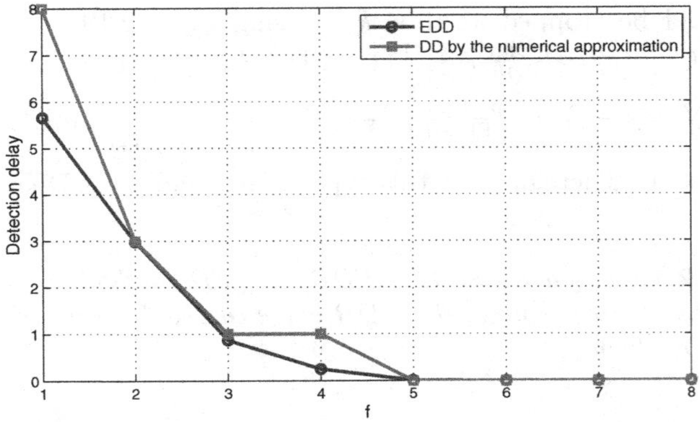

Figure 2.5: Comparison between EDD and the detection delay by numerical approximation for a constant additive fault.

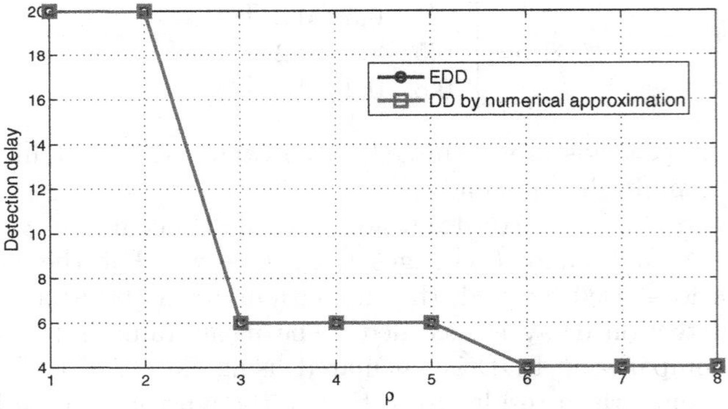

Figure 2.6: Comparison between EDD and the detection delay by numerical approximation for a drift fault.

although the two methods agree with each other, like in the additive fault case, the numerical approximation method contains more fluctuations than the EDD index.

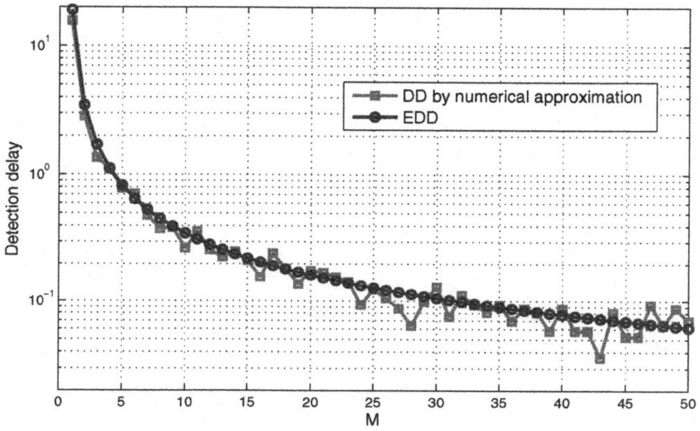

Figure 2.7: Comparison between EDD and the detection delay by numerical approximation for a multiplicative fault.

2.5 Conclusions

This chapter has addressed the fundamental issues for FD of static and dynamic processes. Two types of faults, additive and multiplicative faults have been studied in the way of how they impact the two kinds of processes. To detect such faults, the concept of fault detection statistics was introduced in the statistical framework. In evaluating the performance of FD statistic, the widely accepted performance indices like FDR, FAR have been investigated. In addition, an EDD index was proposed to assess the performance of FD methods for detecting constant additive and multiplicative faults as well as drift faults in stochastic processes. The relationship between EDD and FDR has been also investigated under certain condition. A simple numerical case has been utilized to show the performance of EDD. Compared to numerical approximation-based methods, EDD shows more accurate results.

Figure 2.7 [illegible]

2.5 Conclusions

[illegible]

3 Common test statistics for fault detection

As discussed in Chapter 1, FD test statistics play the central role in PM-FD methods, of which T^2 and Q statistics have been immensely adopted. This chapter will analyse in greater detail their statistical properties to detect additive and multiplicative faults. In addition, due to their lower detectability for the latter type of faults, some alternative statistics will be reviewed and compared.

3.1 Background

Given N measurements, the covariance matrix, Σ_y, can be estimated using $\Sigma_y = \frac{1}{N-1}\sum_{i=1}^{N} \mathbf{y}_i\mathbf{y}_i^T$. To monitor a routine measurement, $\mathbf{y}$, J_{T^2} is, as shown in Section 2.3.1, defined as [2]:

$$J_{T^2} = \mathbf{y}^T\Sigma_y^{-1}\mathbf{y} \tag{3.1}$$

There exist two different detection thresholds for this statistic. The first, often called the empirical threshold, is based on the $\mathcal{F}$-distribution, that is [2, 13]

$$J_{th,T^2} = \frac{m\left(N^2-1\right)}{N\left(N-m\right)}\mathcal{F}_\alpha\left(m, N-m\right) \tag{3.2}$$

The second threshold is based on the χ^2-distribution, that is [2, 13]

$$J_{th,T^2} = \chi^2_{m,\alpha} \tag{3.3}$$

The Q statistic, which is commonly adopted to monitor the residual space, is defined as:

$$J_Q = \mathbf{y}^T\mathbf{y} \tag{3.4}$$

Two methods were proposed to approximate its distribution, namely $\mathrm{tr}(\Sigma_y)\chi_1^2$ and $g\chi_h^2$ with $g = \frac{\mathrm{tr}(\Sigma_y^2)}{\mathrm{tr}(\Sigma_y)}$, $h = \frac{\mathrm{tr}^2(\Sigma_y)}{\mathrm{tr}(\Sigma_y^2)}$ [48], which lead to two types of thresholds. Considering, in addition, methods in [2, 73], there exists a total of four different thresholds for J_Q:

$$J_{th,Q} = \begin{cases} \mathrm{tr}(\Sigma_y)\chi_{1,\alpha}^2 \\ g\chi_{h,\alpha}^2 \\ \lambda_{\max}\chi_{m,\alpha}^2 \\ \theta_1\left[\frac{c_\alpha\sqrt{2\theta_2 h_0^2}}{\theta_1} + 1 + \frac{\theta_2 h_0(h_0-1)}{\theta_1^2}\right]^{\frac{1}{h_0}} \end{cases} \tag{3.5}$$

where $\theta_i = \sum_{j=1}^{m}\sigma_j^i, i = 1,2,3$ with σ_j denoting the jth singular value of Σ_y [73], $h_0 = 1 - (2\theta_1\theta_3)/3\theta_2^2$, c_α is the upper α percentile of the standard normal distribution, and $\lambda_{\max}$ is the maximum eigenvalue of Σ_y.

The multiple choices of threshold are troublesome for user to select a suitable one. Therefore, before examining the detectability of the two statistics, there is a need to investigate their statistical properties in order to determine which threshold to be used under which conditions. It can further be noted that the two methods provide different performance in FDR for additive and multiplicative faults. Such difference should be identified as well.

3.2 Statistical properties of the T^2- and Q-statistics

The following theorem proves the commonly assumed conclusion for Eqs. (3.2) and (3.3):

Theorem 3.1. *Assuming that N is sufficiently large, then both thresholds (Eqs. (3.2) and (3.3)) will give the same result.*

Proof. When N is large, the estimate of the covariance, Σ_y, will be consistent and (almost) equal to the true value. Thus, J_{T^2} will follow a χ_m^2 distribution and $\chi_{m,\alpha}^2$ fits well for the J_{th,T^2}. For the $\mathcal{F}$-based threshold,

$$\lim_{N\to\infty}\frac{m(N^2-1)}{N(N-m)}\mathcal{F}_\alpha(m, N-m) = \lim_{N\to\infty} m\mathcal{F}_\alpha(m,N) = \chi_{m,\alpha}^2 \tag{3.6}$$

Thus, in the case of a large N, the two solutions are equivalent. This can also be verified by examining the statistical property of the two distributions that $\lim_{N\to\infty} \mathrm{E}\left(\frac{m(N^2-1)}{N(N-m)}\mathcal{F}(m, N-m)\right) = \lim_{N\to\infty} \frac{m(N^2-1)}{N(N-m-2)} = m = \mathrm{E}\left(\chi^2_m\right)$ and $\lim_{N\to\infty} \mathrm{Var}\left(\frac{m(N^2-1)}{N(N-m)}\mathcal{F}(m, N-m)\right) = \lim_{N\to\infty} \frac{2m^2(N-2)(N^2-1)^2}{mN^2(N-m-2)^2(N-m-4)} = 2m = \mathrm{Var}\left(\chi^2_m\right)$.

□

In Eq. (3.5), the first 3 thresholds are all based on some modification of the χ^2-distribution. The last threshold is applied to bound PCA-based residual components. Since the focus of this thesis is on the χ^2-distribution, this threshold will not be considered. It should be noted that, unlike J_{T^2}, there is no threshold determination in terms of the exact distribution function for J_Q, while there are only approximate estimates. Thus, it is necessary to consider whether each estimate is biased or not, namely, whether the expectation of each estimate is consistent with the true mean value of J_Q.

Theorem 3.2. *Of the first three thresholds in Eq. (3.5), only the first two are unbiased, while the third one is biased.*

Proof. Before we can prove the above statements, it is necessary to establish the following lemma.

Lemma 1. *Given a quadratic form $x^T\Lambda x$, where x is a vector of the m-dimensional random variables with mean μ and variance Σ, and Λ is an m-dimensional symmetric matrix, then* $\mathrm{E}\left(x^T\Lambda x\right) = \mathrm{tr}\left(\Lambda\Sigma\right) + \mu^T\Lambda\mu$, $\mathrm{Var}\left(x^T\Lambda x\right) = 2\mathrm{tr}\left[\Lambda\Sigma\Lambda\Sigma\right] + 4\mu^T\Lambda\Sigma\Lambda\mu$ *[61, 76].*

From Lemma 1, $\mathrm{E}\left(J_Q\right) = \mathrm{tr}\left(\Sigma_y\right)$ and $\mathrm{Var}\left(J_Q\right) = 2\mathrm{tr}\left(\Sigma_y^2\right)$. For the first threshold, it is obtained that $\mathrm{E}\left(\mathrm{tr}\left(\Sigma_y\right)\chi^2_1\right) = \mathrm{tr}\left(\Sigma_y\right) = \mathrm{E}\left(J_Q\right)$, $\mathrm{Var}\left(\mathrm{tr}\left(\Sigma_y\right)\chi^2_1\right) = 2\mathrm{tr}^2\left(\Sigma_y\right) \geq \mathrm{Var}\left(J_Q\right)$. The second threshold gives $\mathrm{E}\left(g\chi^2_h\right) = \mathrm{tr}(\Sigma_y) = \mathrm{E}\left(J_Q\right), \mathrm{Var}\left(g\chi^2_h\right) = 2\mathrm{tr}(\Sigma_y^2) = \mathrm{Var}\left(J_Q\right)$. Finally, the third threshold gives $\mathrm{E}\left(\lambda_{\max}\chi^2_m\right) = m\lambda_{\max}$ and $\mathrm{Var}\left(\lambda_{\max}\chi^2_m\right) = 2m\lambda^2_{\max}$.

From above results, we can see that the first and second thresholds are unbiased, while the third estimate is biased, since $m\lambda_{max} \geq \mathrm{tr}(\Sigma_y)$. Of the two unbiased thresholds, the first one has a larger variance than the second one. Therefore, using the third threshold will provide conservative bounds on the process. □

Noted that the third threshold was obtained based on $\mathbf{y}^T\mathbf{y} = \left(\mathbf{y}^T\Sigma_y^{-1/2}\right)\Sigma_y\left(\Sigma_y^{-1/2}\mathbf{y}\right) \leq \lambda_{\max}\mathbf{y}^T\Sigma_y^{-1}\mathbf{y}$, then $J_{th,Q} = \lambda_{max}J_{th,T^2} = \lambda_{max}\chi^2_{m,\alpha}$. Thus, this threshold can be set as a benchmark for comparing the other two statistics.

Lemma 2. *For* $x_i \in \mathbb{R}$, $i = 1,...,m$, $\frac{1}{m}(x_1 + x_2,...,+x_m)^2 \leq x_1^2 + x_2^2,...,+x_m^2$ *holds.*

Proof.

$$\begin{aligned}
\frac{1}{m}(x_1 + x_2, \ldots, +x_m)^2 &= \frac{1}{m}\left[\sum_{i=1}^{m} x_i^2 + 2\sum_{i=1}^{m-1}\sum_{j=i+1}^{m} x_i x_j\right] \\
&\leq \frac{1}{m}\left[\sum_{i=1}^{m} x_i^2 + \sum_{i=1}^{m-1}\sum_{j=i+1}^{m}\left(x_i^2 + x_j^2\right)\right] \\
&= \frac{1}{m}\left[\sum_{i=1}^{m} x_i^2 + (m-1)\sum_{i=1}^{m} x_i^2\right] \\
&= x_1^2 + x_2^2 + \ldots + x_m^2
\end{aligned} \tag{3.7}$$

□

Since h can be rewritten as $h = \left(\sum_{i=1}^{m}\lambda_i\right)^2 \Big/ \sum_{i=1}^{m}\lambda_i^2$ with λ_i being the ith eigenvalue of Σ_y, and with the aid of Lemma 2, it is noted that $1 \leq h \leq m$. Considering $g = \frac{\text{tr}\left(\Sigma_y^2\right)}{\text{tr}\left(\Sigma_y\right)} = \frac{\sum_{i=1}^{m}\lambda_i^2}{\sum_{i=1}^{m}\lambda_i} \leq \frac{\lambda_{\max}\sum_{i=1}^{m}\lambda_i}{\sum_{i=1}^{m}\lambda_i} = \lambda_{\max}$, we conclude that $g\chi^2_{h,\alpha} \leq \lambda_{\max}\chi^2_{m,\alpha}$. Theorem 3.2 shows that $\text{tr}(\Sigma_y)\chi_1^2$ gives a larger variance estimation for J_Q, which implies that $g\chi^2_{h,\alpha} \leq \text{tr}\left(\Sigma_y\right)\chi^2_{1,\alpha}$. Note that it is difficult to theoretically compare $\text{tr}\left(\Sigma_y\right)\chi^2_{1,\alpha}$ and $\lambda_{\max}\chi^2_{m,\alpha}$.

Remark 3.1. *The above two statistics define two normal operating region for process data.* J_Q *shows a* m*-dimensional sphere expressed as* $\sum_{i=1}^{m}\mathbf{y}_i^2 \leq J_{th,Q}$, *while* J_{T^2} *gives an* m*-dimensional ellipsoid expressed as*

$\mathbf{y}^T\Sigma_y^{-1}\mathbf{y} \leq J_{th,T^2}$. *Noting that when using* $J_{th,Q} = \lambda_{max}\chi_m^2$, *the ellipsoid will be completely inside the sphere. Thus, if a fault, irrespective of fault type, can be detected by* J_Q, *it will surely be detected by* J_{T^2}. *However, if choosing the other two* $J_{th,Q}$, *the sphere and ellipsoid may be intersected. In this case, comparison of the detectability between them will be troublesome.*

Remark 3.2. *For the case that* Σ_y *is diagonal with the same diagonal element, it gives* $\lambda_i = c\ \forall i$, $h = m$ *and* $g = c$. *Under these cases,* $J_{T^2} = c^{-1}\mathbf{y}^T\mathbf{y} \sim \chi_m^2$ *and* $J_Q = \mathbf{y}^T\mathbf{y} \sim c\chi_m^2$, *thus, they are equivalent.*

3.3 Detecting additive faults

From Eq. (2.2), the additive faults are modelled as $\mathbf{y}_f(k) = \mathbf{y}(k) + \Xi f(k)$, k is the time index. When this kind of fault occurs, J_{T^2} changes to $J_{T_f^2}$, that is,

$$J_{T_f^2}(k) = \mathbf{y}_f^T(k)\Sigma_y^{-1}\mathbf{y}_f(k) \sim \chi_m^2\left(\zeta f^2(k)\right) \tag{3.8}$$

where $\zeta = \Xi^T\Sigma_y^{-1}\Xi > 0$. Thus, based on Eq. (2.12), FDR is written as:

$$\begin{aligned} \mathrm{FDR}(k) &= \mathrm{prob}\left(\chi_m^2\left(\zeta f^2(k)\right) > J_{th,T^2} = \chi_{m,\alpha}^2\right) \\ &= 1 - F_{\chi_m^2(\zeta f^2(k))}\left(J_{th,T^2}\right) \end{aligned} \tag{3.9}$$

The result is shown in Figure 3.1, where it can be seen that detecting an additive fault is equivalent to integrating the $J_{T_f^2}$ distribution from J_{th,T^2} to ∞. It should be noted that from Eq. (3.9), FDR $\propto f^2(k)$ holds.

The J_Q-test statistic for additive faults is [72]

$$J_{Q_f}(k) = \mathbf{y}_f^T(k)\mathbf{y}_f(k) \sim \sum_{i=1}^{m}\lambda_i\chi_1^2\left(\delta_i^2(k)\right) \tag{3.10}$$

where $\delta_i(k) = \bar{\Xi}_i f(k)$, $\bar{\Xi} = \mathbf{\Lambda}^{-1/2}\mathbf{P}^T\Xi$, $\mathbf{\Lambda} = \mathrm{diag}(\lambda_1, ..., \lambda_m)$ and $\mathbf{P}$ is the eigenvector of Σ_y. FDR is computed using [62]

$$\begin{aligned} \mathrm{FDR}(k) &= \mathrm{prob}\left(\sum_{i=1}^{m}\lambda_i\chi_1^2\left(\delta_i^2(k)\right) > J_{th,Q}\right) \\ &\approx 1 - \sum_{i=1}^{\infty} c_i F_{\chi_{m+2i}^2}\left(J_{th,Q}/\varrho\right) \end{aligned} \tag{3.11}$$

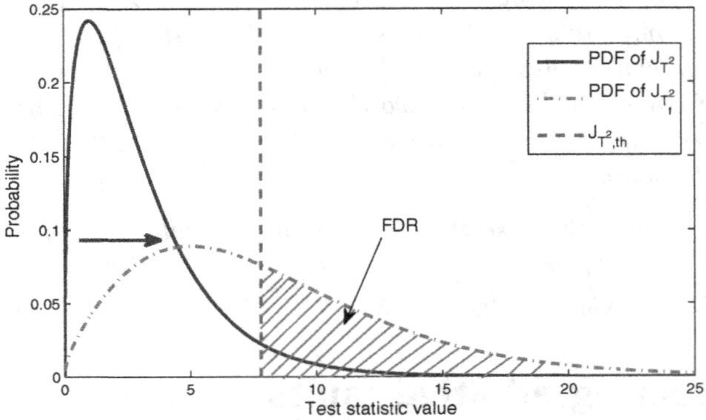

Figure 3.1: Demonstration of J_{T^2} for detecting additive faults

where $\varrho = 0.90625\lambda_{min}$, $c_i \geq 0$ and $\sum_{i=1}^{\infty} c_i = 1$ [62]. Greater details for solving this equation are in [62]. Similarly to J_{T^2}, as $|f(k)|$ increases, FDR(k) will likewise increase. In the case that an additive fault uniformly impacts each variable, *i.e.* $\bar{\Xi}_i = c\ \forall i$, note that

$$\sum_{i=1}^{m} \bar{\Xi}_i^2 = \bar{\Xi}^T\bar{\Xi} = \Xi^T\mathbf{P}\Lambda^{-1}\mathbf{P}^T\Xi = \zeta \tag{3.12}$$

Thus, the FDR of J_{T^2} in this case becomes $\text{FDR}_{T^2} = \text{prob}\left(\chi_m^2\left(m\bar{\Xi}_i^2 f^2\right) > \chi_{m,\alpha}^2\right)$. Since, for a noncentral χ^2-distribution

$$\text{prob}\left(\chi_m^2\left(m\bar{\Xi}_i^2 f^2\right) > \chi_{m,\alpha}^2\right) \geq \text{prob}\left(\chi_1^2\left(\bar{\Xi}_i^2 f^2\right) > \chi_{1,\alpha}^2\right) \tag{3.13}$$

it follows that, since $\text{FDR}_Q = \text{prob}\left(\chi_1^2\left(\bar{\Xi}_i^2 f^2\right) > \chi_{1,\alpha}^2\right)$, $\text{FDR}_{T^2} \geq \text{FDR}_Q$. Thus, it can be concluded that J_{T^2} performs better than J_Q. Above comparison shows a special case, while a general comparison that is difficult to show in the theoretical way will be conducted with a two-dimensional numerical example in Section 3.6.

Remark 3.3. *Ding [2] has deduced that, without a priori fault knowledge (i.e. Ξ), J_{T^2} is equivalent to the generalized likelihood ratio (GLR) test for detecting additive faults. The conclusion, from another viewpoint,* *highlights the uniqueness of J_{T^2} for detecting such faults. Furthermore, by maximizing the likelihood ratio, the fault vector,* $\mathbf{f}$, *is obtained as* $\mathbf{f} = \frac{1}{N_f} \sum_{i=k}^{N_f} \mathbf{y}_f(k)$ *using N_f faulty data.*

Remark 3.4. *If the fault direction vector Ξ is known by a priori or can be estimated from faulty data [13, 14, 23], the two test statistics provide two ways for the online estimation of the fault magnitude in the additive fault case. They are indeed equivalent to the two classic linear estimation methods. Assuming that $\mathbf{y}^* = \mathbf{y}_f - \Xi f$, where $\mathbf{y}^*$ is the fault-free measurements, the method based on J_{T^2} involves the following optimization problem:*

$$\min_f J_{T^2_{f*}(k)} = (\mathbf{y}_f(k) - \Xi f(k))^T \Sigma_y^{-1} (\mathbf{y}_f(k) - \Xi f(k)) \tag{3.14}$$

From Figure 2.1, it is seen that this involves projecting the data back onto the elliptical normal region along Ξ. Solving Eq. (3.14) gives

$$\hat{f}(k) = \left(\Xi^T \Sigma_y^{-1} \Xi\right)^{-1} \Xi^T \Sigma_y^{-1} \mathbf{y}_f(k) \tag{3.15}$$

To estimate $f(k)$, J_Q test statistic seeks to solve the objective function

$$\min_f J_{Q_{f*}(k)} = (\mathbf{y}_f - \Xi f(k))^T (\mathbf{y}_f - \Xi f(k)) \tag{3.16}$$

Similarly, Eq. (3.16) is equivalent to take the abnormal data back to the circular normal region(e.g., those in Figure 3.4 in Section 3.6.1) along Ξ. Solving this optimization gives

$$\hat{f}(k) = \left(\Xi^T \Xi\right)^{-1} \Xi^T \mathbf{y}_f(k) \tag{3.17}$$

Eqs. (3.15) and (3.17) are equivalent to the generalized LS and LS-based estimates of f, respectively. Both methods give an unbiased estimate of f, while J_Q-based one shows a constant estimation variance: $\hat{f} - f \sim \mathcal{N}(0, 1)$, and J_{T^2}-based one gives $\hat{f} - f \sim \mathcal{N}\left(0, \left(\Xi^T \Sigma_y^{-1} \Xi\right)^{-1}\right)$. The first method that depends on the actual variable covariance information, is more flexible and appealing in practice [13, 58, 59].

3.4 Detecting independent multiplicative faults

From the previous definition, the independent multiplicative faults is written as $\mathbf{y}_f = \mathbf{My}$, with $\mathbf{M}$ being a diagonal matrix. In presence of such faults, using the approach obtaining the second estimate of J_Q in Eq. (3.5), J_{T^2} will change to

$$J_{T_f^2} = \mathbf{y}^T \mathbf{M} \Sigma_y^{-1} \mathbf{My} \sim g_f \chi^2_{h_f} \tag{3.18}$$

where $g_f = \frac{\mathrm{tr}(\Sigma_y^{-1}\mathbf{M}\Sigma_y\mathbf{M}\Sigma_y^{-1}\mathbf{M}\Sigma_y\mathbf{M})}{\mathrm{tr}(\Sigma_y^{-1}\mathbf{M}\Sigma_y\mathbf{M})}$ and $h_f = \frac{\mathrm{tr}^2(\Sigma_y^{-1}\mathbf{M}\Sigma_y\mathbf{M})}{\mathrm{tr}(\Sigma_y^{-1}\mathbf{M}\Sigma_y\mathbf{M}\Sigma_y^{-1}\mathbf{M}\Sigma_y\mathbf{M})}$. Two theorems are presented to show the bounds of g_f and h_f. Before giving these theorems, it is first necessary to establish two lemmata that will allow the proofs to be more easily followed.

Lemma 3. *For any $\mathcal{A} \in \mathbb{R}^{n \times n}$ matrix, $\mathrm{tr}(\mathcal{A}\mathcal{A}^T) \geq \mathrm{tr}(\mathcal{A}^2)$ holds.*

Proof. See [75] for a detailed proof. □

Lemma 4. *Given a square orthogonal matrix $\mathfrak{B} \in \mathbb{R}^{n \times n}$, if the eigenvalues happen to be real, then they are forced to be ± 1.*

Proof. Let $\mathbf{v} \in \mathbb{R}^n$ and $\lambda \in \mathbb{R}$ denote the eigenvector and eigenvalue of $\mathfrak{B}$, then $\mathfrak{B}\mathbf{v} = \lambda \mathbf{v}$. Note that

$$\lambda^2 \mathbf{v}^T \mathbf{v} = \mathbf{v}^T \mathfrak{B}^T \mathfrak{B} \mathbf{v} = \mathbf{v}^T \mathbf{v} \tag{3.19}$$

which gives $\lambda^2 = 1$. As it is assumed to be real, λ must be 1 or -1, which establishes the lemma. □

Theorem 3.3. *If $M_i \geq 1$ $\forall i$, then $g_f \geq 1$ holds. Furthermore, given $M_i \geq 1$, g_f becomes 1 if and only if $M_i = 1$ $\forall i$.*

Proof. Noting that $\mathrm{tr}(\mathcal{C}\mathcal{C}^T) = \|\mathcal{C}\|_{\mathrm{F}}^2$ [74], the denominator of g_f is simplified to

$$\begin{aligned}
&\mathrm{tr}\left(\Sigma_y^{-1}\mathbf{M}\Sigma_y\mathbf{M}\right) \\
&= \mathrm{tr}\left(\Sigma_y^{-1/2}\mathbf{M}\Sigma_y\mathbf{M}\Sigma_y^{-1/2}\right) \\
&= \mathrm{tr}\left(\underbrace{\Sigma_y^{-1/2}\mathbf{M}\Sigma_y^{1/2}}_{\mathcal{C}}\underbrace{\Sigma_y^{1/2}\mathbf{M}\Sigma_y^{-1/2}}_{\mathcal{C}^T}\right) \\
&= \|\mathcal{C}\|_{\mathrm{F}}^2
\end{aligned} \tag{3.20}$$

Similarly, the numerator of g_f is rewritten as

$$\begin{aligned}
&\operatorname{tr}\left(\Sigma_y^{-1}\mathbf{M}\Sigma_y\mathbf{M}\Sigma_y^{-1}\mathbf{M}\Sigma_y\mathbf{M}\right)\\
&=\operatorname{tr}\left(\Sigma_y^{-1/2}\mathbf{M}\Sigma_y\mathbf{M}\Sigma_y^{-1}\mathbf{M}\Sigma_y\mathbf{M}\Sigma_y^{-1/2}\right)\\
&=\operatorname{tr}\left(\underbrace{\Sigma_y^{-1/2}\mathbf{M}\Sigma_y^{1/2}}_{\mathcal{C}}\underbrace{\Sigma_y^{1/2}\mathbf{M}\Sigma_y^{-1/2}}_{\mathcal{C}^T}\underbrace{\Sigma_y^{-1/2}\mathbf{M}\Sigma_y^{1/2}}_{\mathcal{C}}\underbrace{\Sigma_y^{1/2}\mathbf{M}\Sigma_y^{-1/2}}_{\mathcal{C}^T}\right)\\
&=\left\|\mathcal{C}\mathcal{C}^T\right\|_{\mathrm{F}}^2
\end{aligned} \tag{3.21}$$

Since the square of Frobenius norm of a matrix is equal to the sum of its squared singular values, Eqs. (3.20) and (3.21) are calculated as:

$$\|\mathcal{C}\|_{\mathrm{F}}^2=\sum_{i=1}^{m}\sigma_{\mathcal{C},i}^2,\quad \left\|\mathcal{C}\mathcal{C}^T\right\|_{\mathrm{F}}^2=\sum_{i=1}^{m}\sigma_{\mathcal{C},i}^4 \tag{3.22}$$

where $\sigma_{\mathcal{C},i}$ is the ith singular value of $\mathcal{C}$. Applying Lemma 2 gives the following inequality

$$\frac{\left\|\mathcal{C}\mathcal{C}^T\right\|_{\mathrm{F}}^2}{\|\mathcal{C}\|_{\mathrm{F}}^2}=\frac{\sum\limits_{i=1}^{m}\sigma_{\mathcal{C},i}^4}{\sum\limits_{i=1}^{m}\sigma_{\mathcal{C},i}^2}\geq\frac{\sum\limits_{i=1}^{m}\sigma_{\mathcal{C},i}^2}{m} \tag{3.23}$$

It should be noted that $\mathcal{C}$ and M are similar since $\mathcal{C}=\Sigma_y^{-1/2}\mathbf{M}\Sigma_y^{1/2}$. This implies that both have the same eigenvalues *i.e.* M_i, $i=1,...,m$. Thus, applying Lemma 3 gives

$$\sum_{i=1}^{m}\sigma_{\mathcal{C},i}^2=\|\mathcal{C}\|_{\mathrm{F}}^2=\operatorname{tr}\left(\mathcal{C}\mathcal{C}^T\right)\geq\operatorname{tr}\left(\mathcal{C}^2\right)=\sum_{i=1}^{m}M_i^2 \tag{3.24}$$

As a result, the following inequality holds

$$g_f=\frac{\left\|\mathcal{C}\mathcal{C}^T\right\|_{\mathrm{F}}^2}{\|\mathcal{C}\|_{\mathrm{F}}^2}\geq\frac{\sum\limits_{i=1}^{m}\sigma_{\mathcal{C},i}^2}{m}\geq\frac{\sum\limits_{i=1}^{m}M_i^2}{m}\geq\frac{m}{m}=1 \tag{3.25}$$

which proves $g_f\geq 1$. Note that from Eq. (3.25), $g_f=1$ gives $M_i=1\ \forall i$. Conversely, plugging $M_i=1$ back to Eq. (3.18) leads to $g_f=1$. Thus, it is concluded that if and only if $M_i=1$, g_f becomes 1. □

Theorem 3.4. *Given $M_i \geq 1\ \forall i$, then $1 \leq h_f \leq m$ holds. Furthermore, $h_f = m$ if and only if $M_i = c\ \forall i,\ c \geq 1$.*

Proof. From Eqs. (3.20)-(3.22), h_f is rewritten as

$$h_f = \frac{\left(\mathrm{tr}\left(\Sigma_y^{-1}\mathbf{M}\Sigma_y\mathbf{M}\right)\right)^2}{\mathrm{tr}\left(\Sigma_y^{-1}\mathbf{M}\Sigma_y\mathbf{M}\Sigma_y^{-1}\mathbf{M}\Sigma_y\mathbf{M}\right)} = \frac{\|\mathcal{C}\|_{\mathrm{F}}^4}{\|\mathcal{C}\mathcal{C}^T\|_{\mathrm{F}}^2} = \frac{\left(\sum_{i=1}^{m}\sigma_{\mathcal{C},i}^2\right)^2}{\sum_{i=1}^{m}\sigma_{\mathcal{C},i}^4} \tag{3.26}$$

Similar to the procedure for proving Theorem 3.3, we get

$$1 \leq \frac{\left(\sum_{i=1}^{m}\sigma_{\mathcal{C},i}^2\right)^2}{\sum_{i=1}^{m}\sigma_{\mathcal{C},i}^4} \leq \frac{\left(\sum_{i=1}^{m}\sigma_{\mathcal{C},i}^2\right)^2}{\frac{1}{m}\left(\sum_{i=1}^{m}\sigma_{\mathcal{C},i}^2\right)^2} = m \tag{3.27}$$

This shows that $1 \leq h_f \leq m$. From Eq. (3.27), $h_f = m$ gives $\sigma_{\mathcal{C},i} = c$, $\forall i$, which implies that $\mathcal{C} = c\mathcal{U}\mathcal{V}^T = c\mathcal{Q}$ with $\mathcal{U}$ and $\mathcal{V}$ being the left and right singular vector matrices and $\mathcal{Q} = \mathcal{U}\mathcal{V}^T$ an orthogonal matrix. Note that since $\mathcal{C}$ has been shown to have the eigenvalues $M_i \geq 1$, then based on Lemma 4, performing eigenvalue decomposition on $\mathcal{C}$ gives the eigenvalue matrix $c\mathbf{I}_m$, which shows that $M_i = c$, $\forall i$. Conversely, if $M_i = c$, it is easy to show that $\sigma_{\mathcal{C},i} = c$ and $h_f = m$. Therefore, it can be concluded that if and only if $M_i = c\ \forall i$, $h_f = m$ holds. □

Considering the special case with $M_i = F > 1\ \forall i$, then $\mathrm{M} = F\mathbf{I}_m$, it further leads to $g_f = F^2$, $h_f = m$. FDR for this kind of fault is expressed as:

$$\mathrm{FDR} = \mathrm{prob}\left(\chi_m^2 > \frac{\chi_{m,\alpha}^2}{F^2}\right) \tag{3.28}$$

Based on Eq. (2.12), detection of this fault is shown in Figure 3.2. It is seen from Eq. (3.28) and Figure 3.2 that as F increases, the FDR will likewise increase. This is similar to that what is observed for the additive case. However, there are a few key differences:

- Unlike the additive case shown in Figure 3.1, where the PDF of $J_{T_f^2}$ shifts to the right as the fault magnitude increases, for a multiplicative fault, Figure 3.2 shows that the threshold J_{th,T^2} shifts

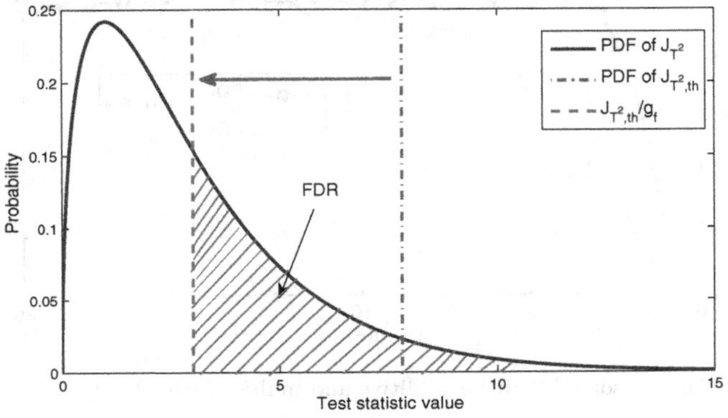

Figure 3.2: Demonstration of J_{T^2} for detecting multiplicative faults

towards the left. Thus, for the multiplicative fault, this is equivalent to integrating the $J_{T_f^2}$ distribution in the interval, $\left[\frac{J_{th,T^2}}{F^2}, \infty\right]$.

- $F \to \infty$ leads to FDR $\to 1$, but FDR $= 1$ is not reachable in practice. This points to a weakness of J_{T^2} for detecting multiplicative faults. Figure 3.3 shows a comparison of the FDR between Eqs. (3.9) and (3.28). It can be seen that for additive faults, as f increases, FDR rapidly reaches 1, while for multiplicative faults, FDR cannot reach 1 even as F approaches infinity.

- For a general multiplicative fault model, where $M_i \neq M_j$, it is possible that $h_f < m$. In this case, there exists a risk that even if $g_f > 1$ this fault cannot be detected due to FDR $< \alpha$.

- From Figure 3.2, it is found that when $F < 1$, $J_{T^2,th}$ will move towards right. In this case, FDR will decrease below α. This case accounts for the fact that $J_{T^2,th}$ cannot detect the fault that decreases the measurement variance.

Given an independent multiplicative fault with $M_i = F \geq 1$. In this case, J_{Q_f} is distributed as $g_{Q,f}\chi^2_{h_{Q,f}}$, where $g_{Q,f} = \frac{\mathrm{tr}(\mathbf{M}\Sigma_y\mathbf{M}\Sigma_y)}{\mathrm{tr}(\mathbf{M}\Sigma_y)} = Fg$

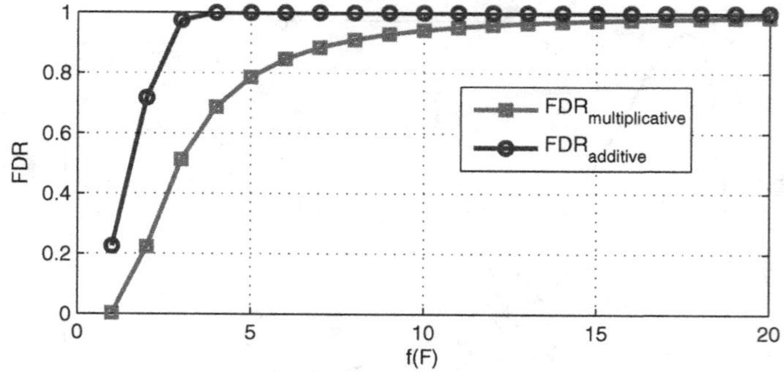

Figure 3.3: Comparison of FDR for additive and multiplicative faults

and $h_{Q,f} = \frac{\mathrm{tr}^2(\Sigma_y \mathbf{M})}{\mathrm{tr}(\mathbf{M}\Sigma_y\mathbf{M}\Sigma_y)} = h$. Then, the FDR is

$$\mathrm{FDR}_Q = \mathrm{prob}\left(\chi_h^2 > \frac{g}{g_{Q,f}}\chi_{h,\alpha}^2 = \frac{\chi_{h,\alpha}^2}{F}\right) \tag{3.29}$$

Since $F \geq 1$, $\mathrm{FDR}_Q \geq \alpha$. Thus, J_Q can detect this fault. Next, note that

$$\mathrm{FDR}_Q = \mathrm{prob}\left(\chi_h^2 > \frac{\chi_{h,\alpha}^2}{F}\right) \leq \mathrm{prob}\left(\chi_h^2 > \frac{\chi_{h,\alpha}^2}{F^2}\right) \tag{3.30}$$

Since $h \leq m$, based on the properties of χ^2gives:

$$\mathrm{prob}\left(\chi_h^2 > \frac{\chi_{h,\alpha}^2}{F^2}\right) \leq \mathrm{prob}\left(\chi_m^2 > \frac{\chi_{m,\alpha}^2}{F^2}\right) = \mathrm{FDR}_{T^2} \tag{3.31}$$

Thus, J_{T^2} performs better than J_Q for this type of fault.

Remark 3.5. *Note that in the general form of* M, *i.e.* $M_i \geq 1$, $g_{Q,f} \geq g$ *and* $\mathrm{h}_{Q,f} \leq h$ *cannot be guaranteed. However, when using* $J_{th,Q} = \mathrm{tr}(\Sigma_y)\chi_{1,\alpha}^2$, *there is no such problem since* $\mathrm{tr}(\mathbf{M}\Sigma_y\mathbf{M}) \geq \mathrm{tr}(\Sigma_y)$ *is always true for* $M_i \geq 1$ $\forall i$. *Thus, based on the above analysis, when* $M_i = F$ $\forall i$ $\mathrm{FDR}_Q = \mathrm{prob}\left(\chi_1^2 > \frac{\chi_{1,\alpha}^2}{F^2}\right) \leq \mathrm{prob}\left(\chi_m^2 > \frac{\chi_{m,\alpha}^2}{F^2}\right) = \mathrm{FDR}_{T^2}$ *is still true.*

Remark 3.6. *Similarly to detect an additive fault, this section provides a comparison for the special multiplicative fault. A more general comparison will be performed in section 3.6, where the numerical simulation is implemented.*

3.5 Alternative statistics for detecting multiplicative faults

As shown in Section 3.4, T^2- and Q-statistics cannot well address multiplicative faults. To mitigate this problem, this section introduces some alternative and advanced methods, including the extension of traditional methods, Wishart distribution-based and information theory-based methods.

3.5.1 The extension of traditional methods

This section presents the extended statistics based on T^2 and Q-statistics.

Cumulative T^2-statistics and cumulative Q-statistics

Assuming that $\mathbf{y} \in \mathbb{R}^m$ is independent and identically distributed, based on n consecutive samples of $\mathbf{y}$, a cumulative T^2 statistic is defined as

$$J_{T_n^2} = \sum_{i=1}^{n} J_{T_i^2} = \sum_{i=1}^{n} \mathbf{y}_i^T \Sigma_y^{-1} \mathbf{y}_i \sim \chi_{nm}^2 \tag{3.32}$$

J_{th,T_n^2} is then given by $\chi^2_{nm,\alpha}$. Let $\mathcal{S} = \sum\limits_{i=1}^{n} \mathbf{y}_i \mathbf{y}_i^T$, then $J_{T_n^2} = \operatorname{tr}\left(\mathcal{S}\Sigma_y^{-1}\right)$. The decision logic is

$$\begin{aligned} &J_{T_n^2} > J_{th,T_n^2} \Rightarrow \text{A multiplicative fault has occurred.} \\ &\text{otherwise} \Rightarrow \text{There is no fault in the process.} \end{aligned} \tag{3.33}$$

The cumulative Q statistic is likewise given by

$$J_{Q_n} = \sum_{i=1}^{n} \mathbf{y}_i^T \mathbf{y}_i = \operatorname{tr}(\mathcal{S}) \sim g\chi_{nh}^2 \tag{3.34}$$

The threshold is $J_{Q_n} = g\chi^2_{h,\alpha}$. Then, the decision logic is

$$\begin{array}{l} J_{Q_n} > J_{th,Q_n} \Rightarrow \text{A multiplicative fault has occurred.} \\ \text{otherwise} \Rightarrow \text{There is no fault in the process.} \end{array} \tag{3.35}$$

Statistical local method

Like Section 3.3, performing singular value decomposition on Σ_y gives $\Sigma_y = \mathbf{P}\mathbf{\Lambda}\mathbf{P}^T$ with $\mathbf{P} = [\mathbf{p}_1, ..., \mathbf{p}_m] \in \mathbb{R}^{m\times m}$ and $\mathbf{\Lambda} = \text{diag}(\lambda_1, ..., \lambda_m)$. Let $\mathbf{h}_i \triangleq \mathbf{y}^T\mathbf{p}_i\mathbf{p}_i^T\mathbf{y} - \lambda_i$ $i = 1, ..., m$, and let $\mathbf{h} = [\mathbf{h}_1, ..., \mathbf{h}_m]^T \in \mathbb{R}^m$. It is noted that $\text{E}(\mathbf{h}_i) = 0$ and $\text{E}(\mathcal{H}) = \mathbf{0}$. Let $\phi(\mathbf{h}_i, K) \triangleq \frac{1}{\sqrt{K}}\sum_{j=1}^{K}\mathbf{h}_{i,j}$, and based on the central limit theorem [42], $\lim_{K\to\infty}\psi_K = [\phi(\mathbf{h}_1, K), ..., \phi(\mathbf{h}_m, K)]^T \sim \mathcal{N}_m(0, \Sigma_\psi)$, where $\Sigma_\psi = \lim_{K\to\infty}(1/K)\sum_{j=1}^{K}\sum_{j=1}^{K}\text{E}\left(\psi_K\psi_K^T\right)$. By selecting a proper k_0, $\psi_{k_0} = [\phi(\mathbf{h}_1, k_0), ..., \phi(\mathbf{h}_m, k_0)]$ is obtained. Finally, using the idea of T^2 statistic, the statistical local method, $J_{\mathcal{L}} = \psi_{k_0}{}^T\Sigma_\psi{}^{-1}\psi_{k_0} \sim \chi^2_m$, was proposed for detecting multiplicative faults that impact Σ_y [42, 63, 64]. Based on $J_{th,\mathcal{L}} = \chi^2_{m,\alpha}$, the decision logic is

$$\begin{array}{l} J_{\mathcal{L}} > J_{th,\mathcal{L}} \Rightarrow \text{A fault that can impact } \Sigma_\text{y} \text{ has occurred.} \\ \text{otherwise} \Rightarrow \text{There is no fault in the process.} \end{array} \tag{3.36}$$

3.5.2 Wishart distribution-based methods

In light of the Wishart distribution [54, 55], $\mathcal{S} \sim \mathcal{W}_m(\Sigma_y, n)$ holds. Comparing $\mathcal{S}$ with Σ_y, it is discovered that, given a sufficiently large n, changes in elements of $\mathcal{S}$ can approximately track changes in Σ_y. In this section, three functions of $\mathcal{S}$, *i.e.* 2-norm, trace and determinant are taken into account for designing test statistics.

The method based on 2-norm of $\mathcal{S}$

Let $\bar{\mathbf{y}} = \Sigma_y^{-1}\mathbf{y}$, then $\bar{\mathbf{y}} \sim \mathcal{N}_m(0, \mathbf{I}_m)$. Define $\bar{\mathcal{S}} = \sum_{i=1}^{n}\bar{\mathbf{y}}_i\bar{\mathbf{y}}_i^T$, then $\bar{\mathcal{S}} \sim \mathcal{W}_m(\mathbf{I}_m, n)$. Given $\gamma = \left\|\frac{\bar{\mathcal{S}}}{m}\right\|_2$, it is noted that if $\frac{n}{m} \geq 1$, $\frac{\gamma-\mu_{nm}}{\sigma_{nm}} \sim TW_1$ [100, 101], where TW_1 stands for the Tracy-Wisdom distribution, $\mu_{nm} =$

$\frac{1}{n}\left(\sqrt{n-1/2}+\sqrt{m-1/2}\right)^2, \sigma_{nm} = \sqrt{\frac{\mu_{nm}}{n}}\left(\frac{1}{\sqrt{n-1/2}}+\frac{1}{\sqrt{m-1/2}}\right)^{1/3}$.
Greater details can be found in [101, 104]. Thus a test statistic $J_\gamma = \frac{\gamma-\mu_{nm}}{\sigma_{nm}}$ is developed for detecting changes in γ, which can reflect some changes in $\mathcal{S}$. The upper and lower thresholds, J_{th,γ_1} and J_{th,γ_2}, is determined using TW_{1,α_1} and TW_{1,α_2} by properly selecting α_1 and α_2 [100, 101]. The decision logic is

$$\begin{array}{l} J_\gamma > J_{th,\gamma_1} \text{ or } J_\gamma < J_{th,\gamma_2} \Rightarrow \text{A multiplicative fault has occurred.} \\ \text{otherwise } \Rightarrow \text{There is no fault in the process.} \end{array} \quad (3.37)$$

The method based on trace of S

Let $\mathcal{T} = \mathrm{tr}(\mathcal{S})$, $\mathcal{T} \sim \sum_{i=1}^{m} \lambda_i \chi_n^2$ holds [105], where λ_i denotes the ith eigenvalue of Σ_y. Thus, $J_{\mathcal{T}} = \mathrm{tr}(\mathcal{S})$ can be used as a test statistic, and $J_{th,\mathcal{T}}$ is calculated as $\sum_{i=1}^{m} \lambda_i \chi_{n,\alpha}^2$. Using $J_{\mathcal{T}}$, a multiplicative fault, $\mathbf{y}_f = \mathbf{M}\mathbf{y}$, leads to $\mathcal{S}_f \sim \mathcal{W}_m(\mathbf{M}\Sigma_y\mathbf{M}^T, n)$ with $\mathcal{S} = \sum_{i=1}^{n} \mathbf{y}_{f,i}\mathbf{y}_{f,i}^T$. Then $J_{\mathcal{T}}$ changes to $J_{\mathcal{T},f} = \sum_{i=1}^{m} \lambda_{f,i}\chi_n^2$, where $\lambda_{f,i}$ is the ith eigenvalue of $\Sigma_{y,f}$ based on Eq. (2.4). FDR for detecting multiplicative faults using this statistic is measured by $\mathrm{FDR}_{Q_n} = \int_{\sum_{i=1}^{m} \lambda_i / \sum_{i=1}^{m} \lambda_{f,i}}^{\infty} f_{\chi_n^2}(x)dx$. The decision logic is

$$\begin{array}{l} J_{\mathcal{T}} > J_{th,\mathcal{T}} \Rightarrow \text{A multiplicative fault has occurred.} \\ \text{otherwise} \Rightarrow \text{There is no fault in the process.} \end{array} \quad (3.38)$$

The method based on determinant of $\mathcal{S}$

Let $\mathcal{D}$ be $\left|\frac{\mathcal{S}}{n}\right|$, then it is obtained that $(2n)^m \frac{\mathcal{D}}{|\Sigma_y|} \sim \prod_{i=0}^{m-1} \chi_{2(n-i)}^2$ [106]. If $J_{\mathcal{D}} = (2n)^m \frac{\mathcal{D}}{|\Sigma_y|}$, it can be used to track changes in Σ_y that increase the determinant of $\mathcal{S}$. Then, $J_{th,\mathcal{D}} = \prod_{i=0}^{m-1} \chi_{2(n-i),\alpha}^2$ is similarly calculated.

The decision logic is

$$\begin{aligned} &J_{\mathcal{D}} > J_{th,\mathcal{D}} \Rightarrow \text{A multiplicative fault has occurred.} \\ &\text{otherwise} \Rightarrow \text{There is no fault in the process.} \end{aligned} \tag{3.39}$$

3.5.3 Information theory-based methods

This section brings together three popular concepts in information and communication areas, and introduce the application of them to detect multiplicative faults.

Conditional entropy-based approach

Given $\mathbf{y} \sim \mathcal{N}_m(0, \Sigma_y)$, the entropy of $\mathbf{y}$ is

$$H(\mathbf{y}) = \frac{1}{2} m \ln(2\pi e) + \frac{1}{2} \ln|\Sigma_y| \tag{3.40}$$

Guerrero-Cusumano [108] gives an alternative formula of $H(y)$:

$$\begin{aligned} H(\mathbf{y}) &= \frac{1}{2} m \ln(2\pi e) + \frac{1}{2}\left(\ln\left|\Sigma_d^2\right| + \ln|P_0|\right) \\ &= \frac{1}{2} m \ln(2\pi e) + \frac{1}{2}\sum_{i=1}^{m} \sigma_i^2 - T(\mathbf{y}) \end{aligned} \tag{3.41}$$

where $T(\mathbf{y})$ is called the mutual information of $\mathbf{y}$ [108], $P_0 = \Sigma_d^{-1/2}\Sigma_y\Sigma_d^{-1/2}$ with Σ_d being the diagonal elements of Σ_y and σ_i^2 being the ith diagonal element. Assume that the correlation measured by $T(\mathbf{y})$ is known, then the estimator of $H(\mathbf{y})$ is

$$\hat{H}(\mathbf{y}) = \frac{1}{2} m \ln(2\pi e) + \frac{1}{2}\sum_{i=1}^{m} s_i^2 - T(\mathbf{y}) \tag{3.42}$$

Let $\delta = H(\mathbf{y}) - \hat{H}(\mathbf{y}) = \sum_{i=1}^{m} \ln\left(\frac{s_i^2}{\sigma_i^2}\right)$ which measures the difference between the estimate and theoretical entropy, s_i^2 is the estimate of σ_i^2 using n samples. Define $E = \sqrt{\frac{n-1}{2m}} \sum_{i=1}^{m} \ln\left(\frac{s_i^2}{\sigma_i^2}\right)$, Yeh *et al.* [107] showed that E is distributed asymptotically as a univariate standard normal distribution. Thus, J_E can be taken as a test statistic to detect changes in the signal entropy caused by multiplicative faults. The upper control threshold $J_{th,E1}$ and lower threshold $J_{th,E2}$ is obtained as [107]

$$\begin{aligned} J_{th,E_1} &= m\sqrt{\tfrac{2(n-1)}{m}}\left[G'\left(\tfrac{n-1}{2}\right) - \ln\left(\tfrac{n-1}{2}\right)\right] \\ &\quad + \mathcal{C}_{\alpha/2} n\sqrt{mG''\left(\tfrac{n-1}{2}\right) + \tfrac{2}{n-1}\mathrm{tr}(P_0 - \mathbf{I}_m)^2} \\ J_{th,E_2} &= m\sqrt{\tfrac{2(n-1)}{m}}\left[G'\left(\tfrac{n-1}{2}\right) - \ln\left(\tfrac{n-1}{2}\right)\right] \\ &\quad - \mathcal{C}_{\alpha/2} n\sqrt{mG''\left(\tfrac{n-1}{2}\right) + \tfrac{2}{n-1}\mathrm{tr}(P_0 - \mathbf{I}_m)^2} \end{aligned}$$

where G' and G'' are, respectively, the first and second derivative of the nature logarithm of the Gamma function [132]. $\mathcal{C}_{\alpha/2}$ is the $1-\alpha/2$ quantile of $\mathcal{N}(0,1)$, which is similar with that mentioned in Eq. (3.5). Using this test statistic, the decision logic is

$$\begin{aligned} &J_E > J_{th,E_1} \text{ or } J_E < J_{th,E_2} \Rightarrow \text{A multiplicative fault has occurred.} \\ &\text{otherwise} \Rightarrow \text{There is no fault in the process.} \end{aligned} \tag{3.43}$$

Multivariate mutual information

For the case that Σ_y is diagonal, using the multivariate mutual information, it was proven that $\mathcal{M} = -\left(n-1-\frac{2m+11}{6}\right)\ln\left(|P_0|\right) \sim \chi^2_{\frac{m(m-1)}{2}}$ [108]. Thus, a test statistic is designed with $J_{\mathcal{M}} = -\left(n-1-\frac{2m+11}{6}\right)\ln\left(|P_0|\right)$. The threshold is $J_{th,\mathcal{M}} = \chi^2_{\frac{m(m-1)}{2},\alpha}$, and the decision logic is

$$\begin{aligned} &J_{\mathcal{M}} > J_{th,\mathcal{M}} \Rightarrow \text{A multiplicative fault has occurred.} \\ &\text{otherwise} \Rightarrow \text{There is no fault in the process.} \end{aligned} \tag{3.44}$$

Kullback-Leibler divergence

Let $\breve{\mathbf{y}} = \mathbf{P}^T\mathbf{y}$, which satisfies $\breve{\mathbf{y}} \sim \mathcal{N}_m(0,\mathbf{\Lambda})$ with $\mathbf{\Lambda}$ being diagonal and defined in Eq. (3.12). Assuming that there is no change occurring in the mean of $\breve{\mathbf{y}}$, it has been shown that given a sufficiently large n, $\mathcal{K} = 2n\mathrm{KL}(\hat{f}_{\breve{\mathbf{y}}}, f_{\breve{\mathbf{y}}}) \sim \chi^2_m$, where $\mathrm{KL}\left(\hat{f}_{\breve{\mathbf{y}}}, f_{\breve{\mathbf{y}}}\right) = \frac{1}{2}\sum_{i=1}^{m}\left\{\ln\left(\frac{\lambda_i}{s_i^2}\right) + \frac{s_i^2}{\lambda_i} - 1\right\}$ [102, 103], $f_{\breve{\mathbf{y}}}$ and $\hat{f}_{\breve{\mathbf{y}}}$ denote the actual and online estimated PDF of $\breve{\mathbf{y}}$ and s_i^2 is the estimate of λ_i using n online samples of y. Defining $J_{\mathcal{K}} =$

Table 3.1: Comparison of different test statistics for multiplicative faults

Symbol	Threshold
$J_{T_n^2} = \sum_{i=1}^{n} \mathbf{y}_i^T \Sigma_y^{-1} \mathbf{y}_i$	$a = 1, b = mn$
$J_{Q_n} = \sum_{i=1}^{n} \mathbf{y}_i^T \mathbf{y}_i$	$a = g,\ b = nh$
$J_{\mathcal{T}} = \mathrm{tr}(\mathcal{S})$	$a = \mathrm{tr}(\Sigma_y),\ b = n$
$J_{\mathcal{M}} = -\left(n - 1 - \frac{2m+11}{6}\right) \ln\left(\lvert P_0 \rvert\right)$	$a = 1,\ b = \frac{m(m-1)}{2}$
$J_{\mathcal{K}} = 2n\mathrm{KL}(\hat{f}_{\breve{\mathbf{y}}}, f_{\breve{\mathbf{y}}})$	$a = 1,\ b = m$
$J_{\mathcal{L}} = \psi_{k_0}^T {\Sigma_\psi}^{-1} \psi_{k_0}$	$a = 1,\ b = m$

$2n\mathrm{KL}(\hat{f}_{\breve{\mathbf{y}}}, f_{\breve{\mathbf{y}}})$ gives the Kullback-Leibler divergence-based statistic for detecting changes in Σ_y. Finally, let $J_{th,\mathcal{K}} = \chi^2_{m,\alpha}$, the decision logic is

$$\begin{aligned} &J_{\mathcal{K}} > J_{th,\mathcal{K}} \Rightarrow \text{A multiplicative fault has occurred.} \\ &\text{otherwise} \Rightarrow \text{There is no fault in the process.} \end{aligned} \tag{3.45}$$

3.5.4 Theoretical comparisons

For methods based on the χ^2 distribution, a brief summary is shown in Table 3.1. The associated thresholds can be given in a generic form, $a\chi^2_{b,\alpha}$. Details of a and b are shown in Table 3.1. we can see that $J_{\mathcal{L}}$ and $J_{\mathcal{K}}$ have the same threshold with J_{T^2}. According to Section 3.2, $g \leq \lambda_{\max} \leq \mathrm{tr}\left(\Sigma_y\right)$ and $1 \leq h \leq m$ hold, which hence give that $J_{\mathcal{T}}$ has the largest a and $J_{T_n^2}$ has the largest b in cases with $n > m$. Noted that not all methods can be summarised using this approach. Some, such as the third Wishart-based method, require the use and construction of complex additional parameters and distributions. The methods based on 2-norm of $\mathcal{S}$ and conditional entropy need to compare J against J_{th_1} and J_{th_2} for making decision.

J_{T^2} and J_Q are calculated by only using the current measurement, which delivers an efficient applicability for detecting additive faults. However, for detecting multiplicative faults, using only current measurements is not sufficient. It has been shown in Section 3.4 that a direct use of them, on the one hand, gives the limited FDR, on the other hand, cannot detect faults that decrease the variable variance. The cumulative extension of J_{T^2} and J_Q can improve their performance in FDR. Based on Eqs. (3.28) and (3.32), $\mathrm{FDR}_{T_n^2} = \int_{\chi^2_{nm,\alpha}/g_f}^{\infty} f_{\chi^2_{nh_f}}(x)dx$ could be calculated for $J_{T_n^2}$. Note that the properties of χ^2 distribution give that for $n \geq 1$, $\mathrm{prob}\left(\chi^2_{h_f} > \frac{\chi^2_{h_f,\alpha}}{g_f}\right) \leq \mathrm{prob}\left(\chi^2_{nh_f} > \frac{\chi^2_{nh_f,\alpha}}{g_f}\right)$. Similarly, for $J_{T_n^2}$, it can be assumed that in cases with h_f close to m or g_f sufficiently large, $\mathrm{FDR}_{T^2} = \mathrm{prob}\left(\chi^2_{h_f} > \frac{\chi^2_{m,\alpha}}{g_f}\right) \leq \mathrm{prob}\left(\chi^2_{nh_f} > \frac{\chi^2_{nm,\alpha}}{g_f}\right) = \mathrm{FDR}_{T_n^2}$. Similarly, there exist cases such as $g_{Q,f} \gg g_f$ leading to $\mathrm{FDR}_{Q_n} = \int_{g\chi^2_{nh,\alpha}/g_{Q,f}}^{\infty} f_{\chi^2_{nh_{Q,f}}}(x)dx \geq \mathrm{FDR}_Q$. These conclusions will be shown in Section 3.6.2 using a numerical example.

Remark 3.7. *While improving the FDR performance, $J_{T_n^2}$ and J_{Q_n} will not increase the FAR level. However, such extensions require more online data for calculations, thus, introduce additional detection delay compared with J_{T^2} and J_Q. Furthermore, in cases with appropriate h_f, $h_{Q,f}$, and g_f, they may perform poorer than J_{T^2} and J_Q.*

Wishart distribution-based approaches are useful when there are n samples available. The methods are aimed to catch the different characteristics of $\mathcal{S}$. Of them, J_γ is developed to monitor changes affecting the 2-norm of $\mathcal{S}$. When a change, $\mathbf{M}$, occurring in Σ_y has affected γ, this multiplicative fault can be detected by J_γ. $J_\mathcal{T}$ is applied to measure changes in the trace of $\mathcal{S}$. Note that only faults that increase the measurement variance can result in $\sum_{i=1}^{m} \lambda_{f,i} = \mathrm{tr}(\Sigma_{y,f}) > \mathrm{tr}(\Sigma_y) = \sum_{i=1}^{m} \lambda_i$, thus $J_\mathcal{T}$ is only able to detect such type of multiplicative fault. The property is consistent with J_{T^2} and J_Q. In addition, comparing $J_\mathcal{T}$ and J_{Q_n}, it is noted that they have the same expression while follow different probability distributions. J_{Q_n} is directly derived from the Q-statistic with the distribution approximated using the scaled χ^2 distribution. $J_\mathcal{T}$ is developed based on the property of Wishart distribution. Noted that $\mathrm{E}(J_{Q_n}) = ngh = n\sum_{i=1}^{m} \lambda_i = \mathrm{E}(J_\mathcal{T})$, and $\mathrm{Var}(J_{Q_n}) = 2ng^2h = 2n\sum_{i=1}^{m} \lambda_i^2 \leq \mathrm{Var}(J_\mathcal{T}) = 2n\left(\sum_{i=1}^{m} \lambda_i\right)^2$. Thus, it is obtained that $J_{th,Q_n} \leq J_{th,\mathcal{T}}$. $J_\mathcal{D}$ is applied to monitor the fault that can increase the determinant of $\mathcal{S}$. As $|\mathcal{S}| = \prod_i \lambda_{s,i}$, $\lambda_{s,i}$ denotes the eigenvalue of $\mathcal{S}$, it can be asserted that faults that disturb J_γ will also result in deviations in $J_{\mathcal{D}}$. Finally, to achieve better performance, it would be helpful to incorporate the three statistics.

It is further noted that $J_{T_n^2}$ can also be interpreted using the Wishart distribution as $J_{T_n^2} = \mathrm{tr}(\mathcal{S}\Sigma_y^{-1}) = \mathrm{tr}(\Sigma_y^{-1/2}\mathcal{S}\Sigma_y^{-1/2})$ holds. According to Eq. (3.32), $\mathrm{E}(J_{T_n^2}) = \mathrm{E}(\chi^2_{nm}) = nm$ and $\mathrm{var}(J_{T_n^2}) = \mathrm{var}(\chi^2_{nm}) = 2nm$. Meanwhile, using the property of Wishart distribution [109], $\Sigma_y^{-1/2}\mathcal{S}\Sigma_y^{-1/2}$ follows $\mathcal{W}_m(\mathbf{I}_m, n)$. Thus, $\mathrm{E}(\mathrm{tr}(\Sigma_y^{-1/2}\mathcal{S}\Sigma_y^{-1/2})) = \mathrm{tr}(\mathrm{E}(\Sigma_y^{-1/2}\mathcal{S}\Sigma_y^{-1/2})) = \mathrm{E}(\mathcal{W}_m(\mathrm{I}_m, n)) = nm$, and $\mathrm{Var}(\mathrm{tr}(\Sigma_y^{-1/2}\mathcal{S}\Sigma_y^{-1/2})) = \mathrm{tr}(\mathrm{Var}(\Sigma_y^{-1/2}\mathcal{S}\Sigma_y^{-1/2})) = \mathrm{E}(\mathcal{W}_m(\mathbf{I}_m, n)) = 2nm$. Furthermore note that $J_{\mathcal{T}}$ can be seen as a special case of $J_{T_n^2}$ by letting $\Sigma_y^{-1} = \mathbf{I}_m$.

Compared with the above mentioned methods, J_E, $J_{\mathcal{K}}$, $J_{\mathcal{M}}$ and $J_{\mathcal{L}}$ need relatively more online measurements. This will decrease their efficiency for realtime applications. Of them, J_E, $J_{\mathcal{K}}$ and $J_{\mathcal{L}}$ can deal with faults either decreasing the measurement variances or increasing them. Note that before using $J_{\mathcal{K}}$, $\mathbf{y}$ must be converted to the independent form z. This can, for one thing, increase the computational complexity, for another, it may lose faulty information in the transformation matrix $\mathbf{P}$. $J_{\mathcal{L}}$ suffers the same problem, which only concerns the eigenvalue space, and does not consider the eigenvector space. While compared with J_E and $J_{\mathcal{K}}$, $J_{\mathcal{L}}$ convert the detection of multiplicative faults to the problem of resolving additive faults by using the statistical local approach and central limit theorem. This has been proved to be efficient for faults decreasing the measurement variance. Meanwhile, a moving window method is always incorporated with $J_{\mathcal{L}}$ so as to improve the online applicability. Finally, $J_{\mathcal{M}}$ seems less efficient since it is only efficient for handling faults arising from the measurement variance. Even worse, before using it one has to guarantee a positive definite P_0. Next, we show the equivalence of $J_{\mathcal{L}}$ and $J_{\mathcal{K}}$.

Firstly, in the case with a single variable, namely $m = 1$, $J_{\mathcal{K}}$ is rewritten as

$$J_{\mathcal{K}} = n\left(\ln\left(\frac{1}{x}\right) + (x) - 1\right) \tag{3.46}$$

where $x = \frac{s^2}{\lambda}$. Note that $s^2 \approx \lambda$, namely $x \approx 1$ holds given a sufficiently large n. Thus, $J_{\mathcal{K}}$ can be approximated using the second order Taylor series at $x = 1$ as [67]

$$J_{\mathcal{K}} \approx n\left(\frac{1}{2}(x-1)^2\right) \tag{3.47}$$

Plugging $x = \frac{s^2}{\lambda}$ into Eq. (3.47) leads to

$$\begin{aligned} J_{\mathcal{K}} &\approx n\left(\frac{1}{2}\left(\frac{s^2}{\lambda}-1\right)^2\right) = n\left(\frac{(s^2-\lambda)^2}{2\lambda^2}\right) \\ &= n\left(\frac{\left(\frac{1}{n-1}\sum_{i=1}^{n}\breve{\mathbf{y}}_i^2-\lambda\right)^2}{2\lambda^2}\right) = \frac{n}{(n-1)^2}\left(\frac{\left(\sum_{i=1}^{n}\breve{\mathbf{y}}_i^2-\lambda\right)^2}{2\lambda^2}\right) \\ &\approx \frac{1}{n}\left(\frac{\left(\sum_{i=1}^{n}\breve{\mathbf{y}}_i^2-\lambda\right)^2}{2\lambda^2}\right) = J_{\mathcal{L}} \sim \chi_1^2 \end{aligned} \tag{3.48}$$

Therefore, given a considerably large n, $J_{\mathcal{K}}$ is equivalent to $J_{\mathcal{L}}$. The conclusion is easily extended to the cases with multiple variables, and can also explain the same threshold for the two methods shown in Table 3.1.

Remark 3.8. *Of these test statistics, J_E, $J_{\mathcal{K}}$ and $J_{\mathcal{M}}$ are not real-time applicable, as they are calculated based on a large number of on-line measurements. To achieve an acceptable real-time application, the other methods introduced above are preferably concerned in practice and will be evaluated using numerical cases in the next section.*

3.6 Simulation results

3.6.1 Additive faults

In order to illustrate the results obtained in Sections 3.2 and 3.3, a simulation example will be considered and the achieved results are compared.

For the purposes of the simulation, a two-dimensional numerical case is applied to demonstrate the two test statistics and thresholds. Two hundred random samples are generated from $\mathcal{N}_2(\mathbf{0}, \Sigma_y)$ with $\Sigma_y = \begin{bmatrix} 2 & 0.5 \\ 0.5 & 1 \end{bmatrix}$. Both additive and multiplicative faults will be considered.

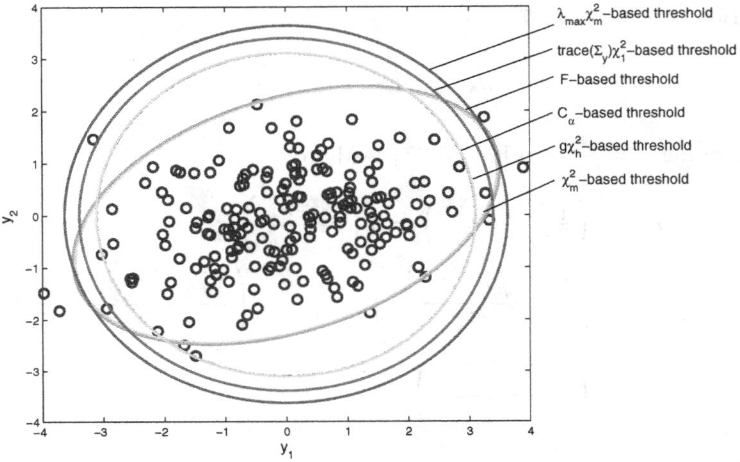

Figure 3.4: Different thresholds for J_{T^2} and J_Q

Figure 3.4 shows the resulting data and the six thresholds obtained using Eqs. (3.2), (3.3) and (3.5). These thresholds defined six different normal data regions, shown as either an ellipse or circle. We can see that the J_{T^2}-based methods produce elliptical regions that can more accurately fit the given data. On the other hand, J_Q-based methods produced circular regions with varying diameters. More specifically, the two J_{th,T^2}-thresholds are similar in shape and size. The $\lambda_{\max}\chi^2_{m,\alpha}$ gives the maximum diameter such that it can completely cover the area contained by J_{th,T^2}. Note that for two-dimensional examples the size of the major axis of the ellipse spanned by J_{T^2} equals to $\lambda_{max}\chi^2_{m,\alpha}$. This shows that the conclusion drawn in [2, 129] that faults that can be detected using J_Q (with $J_{th,Q} = \lambda_{max}\chi^2_{m,\alpha}$) will surely be detected by J_{T^2} (with $J_{th,T^2} = \chi^2_{m,\alpha}$). The $g\chi^2_{h,\alpha}$-bound shows the smallest normal region and acceptable false alarm numbers. The C_α-based method provides similar threshold to $g\chi^2_{h,\alpha}$.

From Section 3.3, solving $\bar{\Xi}_1 = \bar{\Xi}_2$ with constraint $\|\Xi\| = 1$ gives $\Xi = \begin{bmatrix} \pm 0.7429 & \pm 0.6694 \end{bmatrix}^T$ and $\Xi = \begin{bmatrix} \pm 0.9710 & \pm 0.2392 \end{bmatrix}^T$. The four direction vectors are shown in Figure 3.5. It is seen that the two lines approximately cross at the centre point of the circle and ellipses, which provides the boundary of Ξ. When an additive fault occurs in the region going counterclockwise from $\Xi = \begin{bmatrix} \pm 0.9710 & \pm 0.2392 \end{bmatrix}^T$ to $\Xi = \begin{bmatrix} \pm 0.7429 & \pm 0.6694 \end{bmatrix}^T$ irrespective of the fault magnitude, J_{T^2} will be better than J_Q. Two extreme cases are shown corresponding to ($\Xi = \begin{bmatrix} -0.3827 & 0.9239 \end{bmatrix}^T$ and $\Xi = \begin{bmatrix} 0.3827 & -0.9239 \end{bmatrix}^T$), which is resolved by letting $\bar{\Xi}_i = 0$. It can be observed that faults that occur in these directions will be much easier to detect using J_{T2} than J_Q. On the other hand, the region going clockwise is smaller. For faults occurring along directions in this region, it is unclear how to determine the performance of the two statistics. As it is always the case that the process data are correlated with each other, the first region will be bigger than the second one. Thus, in most cases J_{T^2} will perform better than J_Q. Note that for cases with $\Sigma_y = \begin{bmatrix} c & 0 \\ 0 & c \end{bmatrix}$, they are completely equivalent. Table 3.2 shows the comparison using the above numerical case, where the better results are highlighted in red. It is observed that in the first and second cases, J_{T^2} performs better than J_Q, which is consistent with the above conclusion. Note that in the third case (Ξ is along the direction orthogonal to $\Xi = \begin{bmatrix} -0.3827 & 0.9239 \end{bmatrix}^T$), when $f = 1$, and 2 J_{T^2} shows higher FDR, while for $f = 3$, to 5 J_Q is better. This result also agrees with the theory.

Remark 3.9. *In calculating* FDR_Q, $J_{th,Q} = \mathrm{tr}(\Sigma_y)\chi^2_{1,\alpha}$ *is used. This setting allows an easier study on the distribution of* J_Q *under an additive fault. It will be similar when using* $J_{th,Q} = g\chi^2_{h,\alpha}$. *It could be observed from the fourth row of Table 3.2 that as* $g\chi^2_{h,\alpha} < \mathrm{tr}(\Sigma_y)\chi^2_{1,\alpha}$, J_Q *in all fault scenarios is better than* J_{T^2}.

Remark 3.10. *The above results can explain the incorrect use of PCA in PM-FD field. The early implementations of PCA-based FD methods assumed that the residual components obtained from the PCA model contains minor process variations, which makes it problematic to calculate the inverse of its covariance matrix due to limited computer resources, and hence applied* J_Q *to monitor this part. However, currently the computation issue has been well addressed. As discussed above, without a priori fault information,* J_{T^2} *performs definitely better than* J_Q, *thus, there is no need to implement PCA-based methods, while a* J_{T^2} *is suffi-*

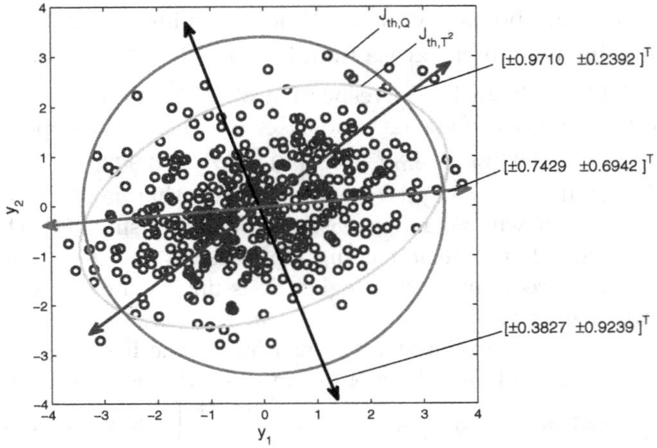

Figure 3.5: Demonstration of J_{T^2} and J_Q for detecting additive faults

Table 3.2: FDR for different additive faults (J_{T^2}/J_Q)

Fault direction	$f=1$	$f=2$	$f=3$	$f=4$
$\Xi = [-0.3827, 0.9239]^T$	0.1563/0.0479	0.5075/0.1585	0.8641/0.4714	0.9859/0.8312
$\Xi = [0.7429, 0.6694]^T$	0.0932/0.0652	0.2421/0.1937	0.4943/0.4285	0.7553/0.6972
$\Xi = [0.9239, 0.3827]^{T\dagger}$	0.0857/0.0672	0.2077/0.1982	0.4225/0.4279	0.6718/0.6867
$\Xi = [0.9239, 0.3827]^{T\ddagger}$	0.0857/0.0933	0.2077/0.2594	0.4225/0.5077	0.6718/0.7538
		$f=5$		
		0.9995/0.9788		
		0.9202/0.8890		
		0.8633/0.8761		
		0.8633/0.9123		

† $J_{th,Q} = \mathrm{tr}(\Sigma_y)\chi^2_{1,\alpha}$
‡ $J_{th,Q} = g\chi^2_{h,\alpha}$.

cient enough. The results also account for why the residual components obtained by PLS model are not appropriate to be monitored using J_Q.

3.6.2 Multiplicative faults

Comparison between J_{T^2} and J_Q

This part further clarifies the results in Sections 3.4 and 3.5. Similar to the additive fault case, when $M_i = F \geq 1 \; \forall i$, the results also define thresholds for multiplicative faults. Figure 3.6 gives three representatives, where the top left shows the multiplicative fault that only occurs in $\mathbf{y}_2$. It is simulated by letting $\Sigma_y = \begin{bmatrix} 2 & 0.5 \\ 0.5 & \boxed{1} \end{bmatrix} \rightarrow \Sigma_y = \begin{bmatrix} 2 & 0.5 \\ 0.5 & \boxed{1.5} \end{bmatrix}$, then $\text{FDR}_{T^2} = 0.1000 \geq \text{FDR}_Q = 0.0688$. we can find that J_{T^2} performs better against J_Q. The top right of Figure 3.6 gives the case discussed above, where $\mathbf{y}_1$ and $\mathbf{y}_2$ are changed by the same M_i value. This fault is simulated by letting $\Sigma_y = \begin{bmatrix} \boxed{2} & 0.5 \\ 0.5 & \boxed{1} \end{bmatrix} \rightarrow \Sigma_y = \begin{bmatrix} \boxed{3} & 0.5 \\ 0.5 & \boxed{1.5} \end{bmatrix}$, then $\text{FDR}_{T^2} = 0.1000 \leq \text{FDR}_Q = 0.1040$. However, for the faults with $M_1 > M_2$, it cannot be guaranteed that FDR_{T^2} is larger than FDR_Q. This extreme case with $M_1 > M_2 = 1$ is shown at the bottom of Figure 3.6, and is simulated by letting $\Sigma_y = \begin{bmatrix} \boxed{2} & 0.5 \\ 0.5 & 1 \end{bmatrix} \rightarrow \Sigma_y = \begin{bmatrix} \boxed{3} & 0.5 \\ 0.5 & 1 \end{bmatrix}$, then $\text{FDR}_{T^2} = 0.1494 \geq \text{FDR}_Q = 0.1246$, thus, it is shown that J_Q performs similarly with J_{T^2}, even better.

Comparisons among alternative statistics

This part examines the methods shown in Section 3.5. To compare the performance of cumulative version of T^2 against the original one presented in Section 3.5.1, we first define the index $\vartheta = \frac{\text{FDR}_{T_n^2}}{\text{FDR}_{T^2}}$. Then, $\vartheta > 1$ means that $J_{T_n^2}$ perform better than J_{T^2}. Assume that there is a process consisting of 10 process variables ($m = 10$) for monitoring. Different multiplicative faults will give different g_f and h_f in J_{T^2} and $J_{T_n^2}$, which thus give different FDR. Figure 3.7 shows the performance of ϑ calculated based on g_f varying from 5 to 10 and h_f varying from 5 to 10. It is shown that in all simulations, ϑ is larger than 1. From Section 3.4, FDR_{T^2} approaches 1 but cannot theoretically reach 1, even though the fault is strong enough. Using $J_{T_n^2}$, in some cases, $\text{FDR}_{T_n^2}$ can easily reach 1. This result is illustrated from the two subfigures of Figure 3.7. The first one takes $g_f = h_f = 5$, it can be observed that $\text{FDR}_{T_n^2} = 1$ compared to $\text{FDR}_{T^2} < 1$. Considering a more severe

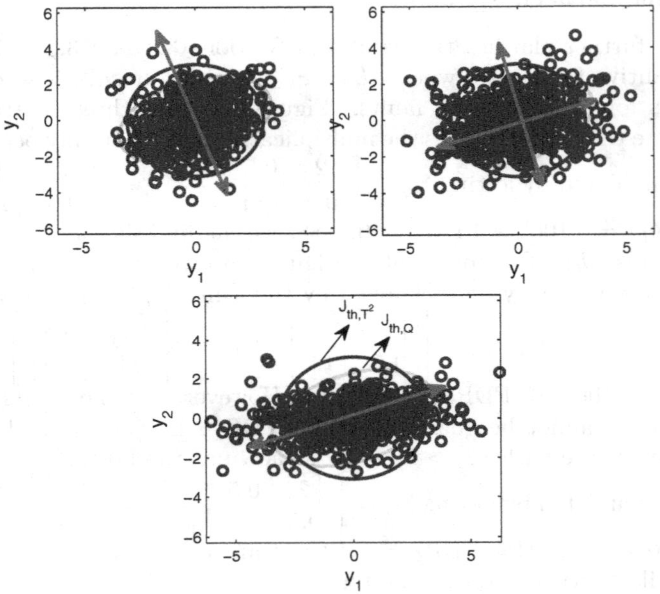

Figure 3.6: Schematic description of J_{T^2} and J_Q for detecting multiplicative faults

fault with $g_f = h_f = 8$, the second subfigure indicates that FDR_{T^2} gets larger and larger, but still less than 1. By contrast, $\mathrm{FDR}_{T_n^2} = 1$ is easily reached. Figure 3.8(a) examines the performance of ϑ with different n, where it is assumed that $m = 10$, $g_f = 5$ and $h_f = 5$. It is seen that when $n = 1$, the two statistics are equivalent, thus $\vartheta = 1$. As n increases, ϑ as well increase up to a stable value. Since, a larger n will increase the online complexity of the statistics, it is practical to choose $n = 8$ or 9 for this fault.

In some cases, $J_{T_n^2}$ performs worse than J_{T^2}. Taking the case with $m = 10$, $g_f = 3$, $h_f = 3$ and $n = 5$ for instance, it gives $\mathrm{FDR}_{T^2} = 0.1067 > 0.0953 = \mathrm{FDR}_{T_n^2}$. Figure 3.8(b) shows the behavior of ϑ with different n. It can be found that in all cases $\vartheta < 1$, and as n increases, ϑ likewise decreases. This proves that when $h_f \ll m$ and g_f is not

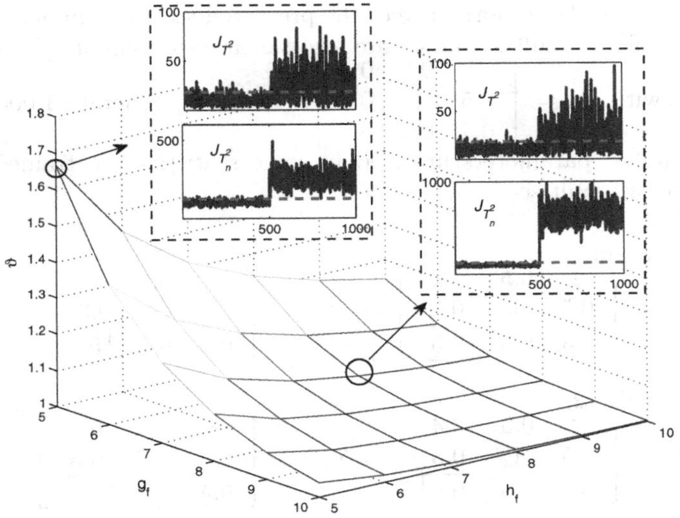

Figure 3.7: Performance of ϑ with different g_f and h_f, $m = 10$, $n = 10$

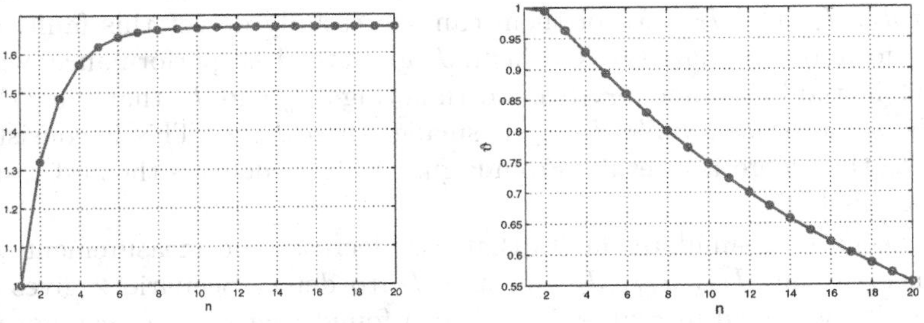

(a) $g_f = 5$ and $h_f = 5$. (b) $g_f = 3$ and $h_f = 3$.

Figure 3.8: Performance of ϑ with different n, $m = 10$

large enough, $J_{T_n^2}$ cannot improve the detection performance even if n is sufficiently large. Note that it is hard to pick up a boundary explicitly stating when $J_{T_n^2}$ performs better than J_{T^2}.

To evaluate the performance of the presented statistics in Section 3.5, some simulated fault scenarios are also taken into account. Consider a process with $\Sigma_y = \begin{bmatrix} 2 & 0.5 & 0.4 \\ 0.5 & 1 & 0.3 \\ 0.4 & 0.3 & 0.5 \end{bmatrix}$. Using Σ_y generate 1,000 fault-free samples, parameters involved in some statistics are trained. We consider two fault scenarios:

- Scenario 1:

$$\Sigma_y = \begin{bmatrix} 2 & 0.5 & 0.4 \\ 0.5 & 1 & 0.3 \\ 0.4 & 0.3 & \boxed{0.5} \end{bmatrix} \to \Sigma_{y,f} = \begin{bmatrix} 2 & 0.5 & 0.4 \\ 0.5 & 1 & 0.3 \\ 0.4 & 0.3 & \boxed{5} \end{bmatrix}.$$

- Scenario 2:

$$\Sigma_y = \begin{bmatrix} 2 & 0.5 & 0.4 \\ 0.5 & 1 & 0.3 \\ 0.4 & 0.3 & \boxed{0.5} \end{bmatrix} \to \Sigma_{y,f} = \begin{bmatrix} 2 & 0.5 & 0.4 \\ 0.5 & 1 & 0.3 \\ 0.4 & 0.3 & \boxed{0.1} \end{bmatrix}$$

For Scenario 1, 1,000 samples are collected where the fault occurs from 501 to 1000. Figure 3.9 shows the performance of J_{T^2}, $J_{T_n^2}$, J_Q and J_{Q_n}. It can be observed that $J_{T_n^2}$ and J_{Q_n} significantly improved the detection compared against J_{T^2} and J_Q. Figure 3.10 shows the results of J_γ, $J_\mathcal{T}$ and $J_\mathcal{D}$. All of them can successfully detect this fault, and perform better than J_{T^2}. Of them, J_γ gives the best performance, which is followed by $J_\mathcal{D}$ and $J_\mathcal{T}$. It is worth noting J_{Q_n} and $J_\mathcal{T}$ that, they show the same profile, while J_{th,Q_n} is smaller than $J_{th,\mathcal{T}}$. This is consistent with the theoretical result. From Figure 3.11, $J_\mathcal{L}$ delivers the best results for this fault.

Scenario 2 simulates faults that can decrease the measurement variance. Using J_{T^2}, $J_{T_n^2}$, $J_\mathcal{D}$, J_γ and $J_\mathcal{L}$ to detect Scenario 2 gives the results as shown in Figure 3.12. It can found that only $J_\mathcal{L}$ catches this fault. Thus, for faults decreasing the measurement variance, $J_\mathcal{L}$ is suggested.

Remark 3.11. *In practice, such changes are always treated as a good sign of operating performance, while in some cases, they are referred to as a fault. For example a sensor malfunctions which leads to the measurements fixed to a constant value, then their variances will decrease to zero.*

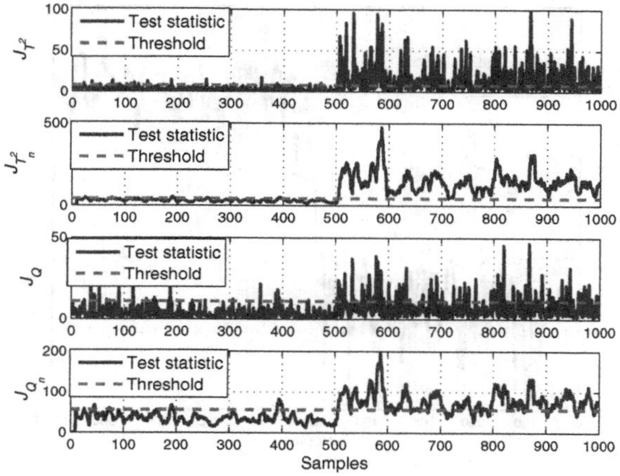

Figure 3.9: Performance of J_{T^2}, $J_{T_n^2}$, J_Q and J_{Q_n} for Scenario 1

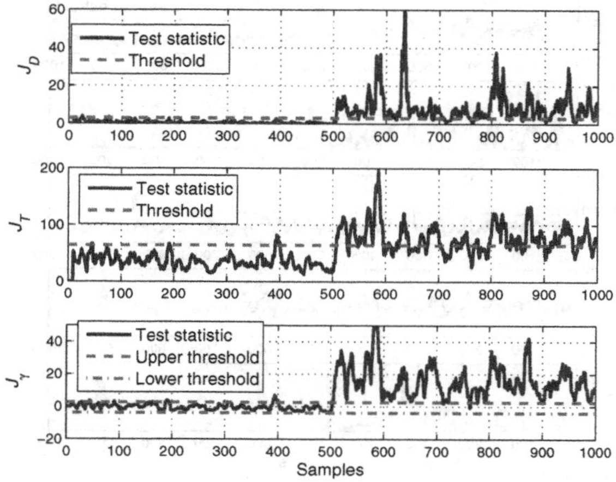

Figure 3.10: Performance of J_γ, $J_{\mathcal{T}}$ and $J_{\mathcal{D}}$ for Scenario 1

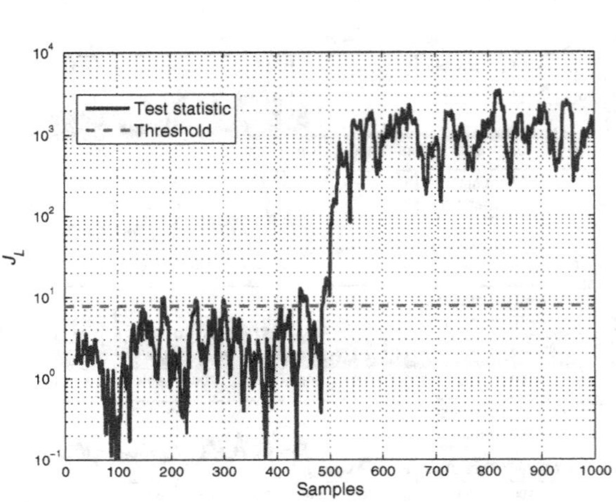

Figure 3.11: Performance of $J_{\mathcal{L}}$ for Scenario 1

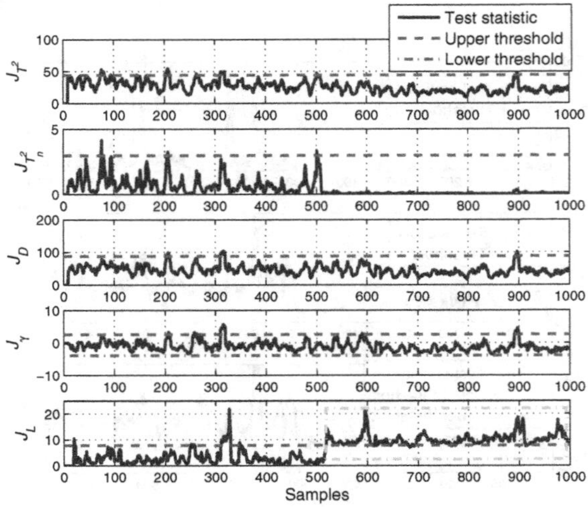

Figure 3.12: Performance of different statistics for Scenario 2

3.7 Conclusions

In the statistical framework, this chapter has firstly examined the ability of the T^2- and Q-test statistics to detect additive and multiplicative faults in terms of the FDR index. Four major conclusions can be drawn: (1) J_{T^2} contains explicit distribution information and, thus, a unique threshold is obtained, while J_Q has only an approximate probability distribution that will give different thresholds; (2) For both statistics, it has been shown that detecting additive faults is equivalent to shifting the PDF of χ^2 to the right, which is achieved by increasing the noncentrality parameter, while detecting multiplicative faults is equivalent to shifting the threshold to the left; (3) In most cases, when detecting additive faults, J_{T^2} performs better than J_Q. In few cases that depend on properties of covariance matrices and the choice of $J_{th,Q}$, J_Q performs better; (4) When detecting independent multiplicative faults, both methods cannot reach FDR $= 1$. Similarly to additive faults, J_{T^2} performs better in most cases. For faults that affect variables with a large variance, the two methods give similar FDR values; although J_Q performs better.

To address the inefficiency when detecting multiplicative faults, some alternative statistics have been reviewed and compared. For this part, it has been drawn that in some cases, $J_{T^2_n}$ as an extension of J_{T^2} can improve the performance of J_{T^2} for faults increasing the measurement variance. While in some other cases, $J_{T^2_n}$ cannot or even decrease the performance of FDR. J_T, J_γ, $J_\mathcal{D}$ and $J_\mathcal{L}$ show a better FDR compared with J_{T^2} and J_Q. It was also found that $J_\mathcal{L}$ is the only method which behaves well for multiplicative fault shrinking the measurement variance. Finally, we would like to further note that the considered statistics can be combined with typical multivariate analysis-based methods, such as PCA, PLS (Chapter 4) and canonical correlation analysis (CCA), *etc.* methods so as to detect faults for static processes, with dynamic PCA, PLS (Chapter 5) and CCA *etc.* models for addressing the FD issue for steady-state dynamic processes and with the data-driven PS- and DO-based methods to deal with dynamic processes in general cases (Chapter 6).

4 KPI-based PM-FD methods for static processes

Based on the fundamental issue resolved in last two chapters, the following chapters turn to the study of KPI-based PM-FD methods in the framework of Figure 1.2. As introduced in Chapter 1, at the plant level, the KPI variable can either be the output variables of a subprocess or directly associated with the process output and input variables. To distinguish them, the notation $\mathbf{y}_{KPI}$ is commonly employed to represent KPI variables, while for the process variables consisting of input $\mathbf{u}$ and output $\mathbf{y}$, $\mathbf{y}_{proc}$ is adopted. For the sake of simplicity, in this thesis, $\mathbf{y}$ and $\boldsymbol{\theta}$ are, respectively, used to denote the readily available process variables and online unmeasurable KPI variables [6]. Additionally, as shown in Chapter 2, $\mathbf{y}_{obs}$ and $\boldsymbol{\theta}_{obs}$ are also adopted to present the raw process and KPI measurements when the data normalization is involved in examined methods. Based on them, various KPI-based PM-FD methods were developed for static and dynamic processes. In this chapter, a thorough comparison study on KPI-based PM-FD methods for static processes, namely the MSPM methods, are presented. Theoretical comparisons in terms of their computations, geometric properties and interconnections will be considered in the first part. Then, their performance to detect constant additive, drift and multiplicative faults is evaluated using the EDD index.

4.1 Background

In the framework of commonly used MSPM methods, the process statistical properties are presented in form of the historical process datasets. From the recorded off-line process data $\mathbf{y}_{obs,i} \in \mathbb{R}^m$ and key performance indicators $\boldsymbol{\theta}_{obs,i} \in \mathbb{R}^l$ with $i = 1, ..., N$, the variable matrices could be formulated in the form of $\mathbf{Y}_{obs} = [\mathbf{y}_{obs,1}, \mathbf{y}_{obs,2}, ..., \mathbf{y}_{obs,N}]$

and $\mathbf{\Theta}_{obs} = [\boldsymbol{\theta}_{obs,1}, \boldsymbol{\theta}_{obs,2}, ..., \boldsymbol{\theta}_{obs,N}]$. Considering the static processes, it is assumed that $\mathbf{y}_{obs} \sim \mathcal{N}_m (\mathrm{E}(\mathbf{y}_{obs}), \mathrm{Cov}(\mathbf{y}_{obs}))$, and $\boldsymbol{\theta}_{obs} \sim \mathcal{N}_l (\mathrm{E}(\boldsymbol{\theta}_{obs}), \mathrm{Cov}(\boldsymbol{\theta}_{obs}))$, where $\mathrm{E}(\mathbf{y}_{obs}) \approx (1/N) \sum_{i=1}^{N} \mathbf{y}_{obs,i}$, $\mathrm{E}(\boldsymbol{\theta}_{obs}) \approx (1/N) \sum_{i=1}^{N} \boldsymbol{\theta}_{obs,i}$. Among MSPM approaches, both the process and KPI data should be mean-centered and normalized, that is

$$\begin{aligned} \mathbf{y} &= \mathrm{diag}\left(\sigma_{y,1}^{-1}, ..., \sigma_{y,m}^{-1}\right) (\mathbf{y}_{obs} - \mathrm{E}(\mathbf{y}_{obs})) \\ \boldsymbol{\theta} &= \mathrm{diag}\left(\sigma_{\theta,1}^{-1}, ..., \sigma_{\theta,l}^{-1}\right) (\boldsymbol{\theta}_{obs} - \mathrm{E}(\boldsymbol{\theta}_{obs})) \end{aligned} \tag{4.1}$$

where $\sigma_{y,j} = \sqrt{\frac{1}{N-1} \sum_{i=1}^{N} (\mathbf{y}_{obs,i}(j) - \mathbf{E}(\mathbf{y}_{obs}(j)))^2}$ for $j = 1, ..., m$, $\sigma_{\theta,k} = \sqrt{\frac{1}{N-1} \sum_{i=1}^{N} (\boldsymbol{\theta}_{obs,i}(k) - \mathbf{E}(\boldsymbol{\theta}_{obs}(k)))^2}$ for $k = 1, ..., l$. Based on the off-line data matrices $\mathbf{Y} = [\mathbf{y}_1, ..., \mathbf{y}_N]$ and $\mathbf{\Theta} = [\boldsymbol{\theta}_1, ..., \boldsymbol{\theta}_N]$, the problem is how to monitor the process performance, in particular the KPI, by only using the normalized online process measurement $\mathbf{y}_{on}$.

An intuitive solution for this problem is to directly design the T^2 statistic introduced in Chapter 3 [52, 54]:

$$T^2 = \mathbf{y}_{on}^T \Sigma_y^{-1} \mathbf{y}_{on} \tag{4.2}$$

and based on the study in Chapter 3, it follows the χ_m^2 distribution in the case of sufficient training samples, otherwise $\frac{m(N^2-1)}{N(N-m)} \mathcal{F}_{m,N-m}$ [2]. However, there remain two main problems, for one thing, Σ_y may be numerical instability, for the important one, the status of KPI cannot be judged only by the detection result. Therefore, an alterative solution using the linear mapping method is considered by projecting $\mathbf{y}_{on}$ onto two subspaces:

$$\mathbf{y}_{on} = \underbrace{\mathbf{\Pi}_\theta \mathbf{y}_{on}}_{\hat{\mathbf{y}}_{on}(\boldsymbol{\theta}-\text{correlated})} + \underbrace{\mathbf{\Pi}_{\theta^\perp} \mathbf{y}_{on}}_{\tilde{\mathbf{y}}_{on}(\boldsymbol{\theta}-\text{uncorrelated})} \tag{4.3}$$

of which the first part is directly correlated with KPI, while the second one is uncorrelated with KPI. Like Eq. (4.2), two detection indices can be respectively designed on the two subspaces for monitoring KPI-correlated and -uncorrelated parts of $\mathbf{y}$, and the former one can be used to judge the status of KPI. In Eq. (4.3), two projectors, $\mathbf{\Pi}_\theta$ and $\mathbf{\Pi}_{\theta^\perp}$ are defined to split $\mathbf{y}_{on}$, where $\mathbf{\Pi}_\theta + \mathbf{\Pi}_{\theta\perp} = \mathbf{I}_m$ holds, which shows the two subspaces jointly form the full $\mathbf{y}$-space. Based on the projectors, let the subspaces $\mathbb{S}_\theta$ and $\mathbb{S}_{\theta\perp}$ be the $\boldsymbol{\theta}$-correlated and $\boldsymbol{\theta}$-uncorrelated subspaces of $\mathbf{y}$. They should satisfy $\dim\{\mathbb{S}_\theta\} + \dim\{\mathbb{S}_{\theta\perp}\} = \dim\{\mathbb{S}_y\} = m$, where $\dim\{.\}$ represents the dimension of a subspace. In [29], the projection property of PLS has been revealed, which motivates us to investigate geometric properties of other methods in this area. Furthermore, their interconnections, inherent differences, and some unique characteristic-s are still confusing. Therefore, the first issue to be addressed in this chapter is the review of existing methods. It includes a comprehensive comparison on the geometric properties of $\mathbf{\Pi}_\theta$, $\mathbf{\Pi}_{\theta\perp}$, $\mathbb{S}_\theta$ and $\mathbb{S}_{\theta\perp}$ obtained by different methods, and different ways in which the PM-FD indices are developed.

4.2 Classification of existing approaches

4.2.1 A direct method

First, the cross-covariance matrix between $\mathbf{y}$ and $\boldsymbol{\theta}$ can be estimated as: $\frac{\mathbf{\Theta Y}^T}{N-1}$. Performing a SVD on it gives:

$$\frac{\mathbf{\Theta Y}^T}{N-1} = \mathbf{V\Sigma U}^T = \mathbf{V}\left[\mathbf{\Sigma}_{\theta y}, 0\right]\begin{bmatrix}\mathbf{U}_{\theta y}^T \\ \mathbf{U}_{\theta y\perp}^T\end{bmatrix} = \mathbf{V\Sigma}_{\theta y}\mathbf{U}_{\theta y}^T \tag{4.4}$$

Since $l = \operatorname{rank}\left(\mathbf{U}_{\theta y}\mathbf{U}_{\theta y}^T\mathbf{Y}\right) \leq \operatorname{rank}\left(\mathbf{U}_{\theta y}^T\mathbf{Y}\right) \leq l$ and $m - l = \operatorname{rank}\left(\mathbf{U}_{\theta y\perp}\mathbf{U}_{\theta y\perp}^T\mathbf{Y}\right) \leq \operatorname{rank}\left(\mathbf{U}_{\theta y\perp}^T\mathbf{Y}\right) \leq m - l$, it follows that $\operatorname{rank}\left(\mathbf{U}_{\theta y}^T\mathbf{Y}\right) = l$ and $\operatorname{rank}\left(\mathbf{U}_{\theta y\perp}^T\mathbf{Y}\right) = m - l$ hold.

Thus, regarding the two monitoring subspaces, two separate T^2 statistics can be designed below

$$\begin{aligned} T_{\hat{y}}^2 &= \mathbf{y}^T\mathbf{U}_{\theta y}\left(\frac{\mathbf{U}_{\theta y}^T\mathbf{YY}^T\mathbf{U}_{\theta y}}{N-1}\right)^{-1}\mathbf{U}_{\theta y}^T\mathbf{y} \\ T_{\tilde{y}}^2 &= \mathbf{y}^T\mathbf{U}_{\theta y\perp}\left(\frac{\mathbf{U}_{\theta y\perp}^T\mathbf{YY}^T\mathbf{U}_{\theta y\perp}}{N-1}\right)^{-1}\mathbf{U}_{\theta y\perp}^T\mathbf{y} \end{aligned} \tag{4.5}$$

Furthermore, it is noted that the inverse calculation may not be stable, since the two covariance matrices may be numerically sensitive.

To overcome this disadvantage, two robust detection statistics are designed as [2, 9]

$$T_{\hat{y}}^2 = \mathbf{y}^T \mathbf{U}_{\theta y} \mathbf{\Xi}_1 \mathbf{U}_{\theta y}^T \mathbf{y},\ T_{\tilde{y}}^2 = \mathbf{y}^T \mathbf{U}_{\theta y^\perp} \mathbf{\Xi}_2 \mathbf{U}_{\theta y^\perp}^T \mathbf{y} \tag{4.6}$$

where $\mathbf{\Xi}_1 = \mathbf{P}_1 \mathrm{diag}\left(\frac{\lambda_{1,l}}{\lambda_{1,1}}, ..., \frac{\lambda_{1,l}}{\lambda_{1,l}}\right) \mathbf{P}_1^T$, $\mathbf{P}_1 \mathbf{\Lambda}_1 \mathbf{P}_1^T = \frac{\mathbf{U}_{\theta y}^T \mathbf{Y}\mathbf{Y}^T \mathbf{U}_{\theta y}}{N-1}$; $\mathbf{\Xi}_2 = \mathbf{P}_2 \mathrm{diag}\left(\frac{\lambda_{2,m-l}}{\lambda_{2,1}}, ..., \frac{\lambda_{2,m-l}}{\lambda_{2,m-l}}\right) \mathbf{P}_2^T$, $\mathbf{P}_2 \mathbf{\Lambda}_2 \mathbf{P}_2^T = \frac{\mathbf{U}_{\theta y^\perp}^T \mathbf{Y}\mathbf{Y}^T \mathbf{U}_{\theta y^\perp}}{N-1}$, $\mathbf{\Lambda}_1 = \mathrm{diag}(\lambda_{1,1}, ..., \lambda_{1,l})$, $\mathbf{\Lambda}_2 = \mathrm{diag}(\lambda_{2,1}, ..., \lambda_{2,m-l})$; $\mathbf{\Lambda}_i$ and $\mathbf{P}_i$ are obtained using either SVD or eigenvalue decomposition. Like that introduced in Chapter 3, the thresholds are obtained from the χ^2 distribution as [2, 6]

$$J_{th,T_{\hat{y}}^2} = \lambda_{1,l} \chi_{l,\alpha}^2,\ J_{th,T_{\tilde{y}}^2} = \lambda_{2,m-l} \chi_{m-l,\alpha}^2 \tag{4.7}$$

Therefore, the decision logic is:

$$\begin{cases} T_{\hat{y}}^2 \le J_{th,T_{\hat{y}}^2} \text{ and } T_{\tilde{y}}^2 \le J_{th,T_{\tilde{y}}^2} \Rightarrow \text{fault} - \text{free} \\ T_{\hat{y}}^2 > J_{th,T_{\hat{y}}^2} \Rightarrow \text{KPI} - \text{related fault occurs} \\ T_{\tilde{y}}^2 > J_{th,T_{\tilde{y}}^2} \Rightarrow \text{KPI} - \text{unrelated fault occurs} \end{cases} \tag{4.8}$$

4.2.2 Linear regression-based methods

In this section, two methods based on linear regression are reviewed.

LS regression-based method

LS plays a major role in linear regression field. It predicts the KPI not only based on the cross-covariance between $\mathbf{y}$ and $\boldsymbol{\theta}$ but also the covariance of $\mathbf{y}$. In the general case, the regression model is expressed as [6, 36]:

$$\hat{\boldsymbol{\theta}}_{LS} = \mathbf{\Psi}_{LS} \mathbf{y},\ \mathbf{\Psi}_{LS} = \mathrm{E}\left(\theta \mathbf{y}^T\right) \mathrm{E}\left(\mathbf{y}\mathbf{y}^T\right)^\dagger = \mathbf{\Theta}\mathbf{Y}^T \left(\mathbf{Y}\mathbf{Y}^T\right)^\dagger \tag{4.9}$$

where $\hat{\boldsymbol{\theta}}_{LS}$ represents the LS-prediction of $\boldsymbol{\theta}$, $\left(\mathbf{Y}\mathbf{Y}^T\right)^\dagger = \mathbf{P}_y \mathbf{\Lambda}_y^{-1} \mathbf{P}_y$, $\mathbf{\Lambda}_y$ and $\mathbf{P}_y$ are the non-zero singular values of $\mathbf{Y}\mathbf{Y}^T$ and the corresponding singular vectors, respectively. To complete the objective of monitoring $\boldsymbol{\theta}$ with $\mathbf{y}$, this method basically decomposes $\mathbf{y}$ based on $\hat{\boldsymbol{\theta}}_{LS}$. The projection performed by Yin *et al.*, first projects $\mathbf{y}$ onto the subspace spanned by the columns of $\mathbf{\Psi}_{LS}^T$, which is supposed to be responsible for KPI prediction, then onto its orthogonal complement subspace, which has no contribution to $\hat{\boldsymbol{\theta}}_{LS}$ [40]. The procedure can be realized with a QR decomposition on $\boldsymbol{\Psi}_{LS}^T$:

$$\boldsymbol{\Psi}_{LS}^T = [\mathbf{Q}_1, \mathbf{Q}_2] \begin{bmatrix} \mathbf{R}_1 \\ 0 \end{bmatrix} = \mathbf{Q}_1\mathbf{R}_1 \tag{4.10}$$

Since $\text{rank}\left(\mathbf{Q}_1^T\mathbf{y}\right) = l$ and $\text{rank}\left(\mathbf{Q}_2^T\mathbf{y}\right) = m - l$, two T^2 statistics are designed for them as

$$\begin{aligned} T^2_{\hat{y}_{LS}} &= \mathbf{y}^T\mathbf{Q}_1\left(\frac{\mathbf{Q}_1^T\mathbf{Y}\mathbf{Y}^T\mathbf{Q}_1}{N-1}\right)^{-1}\mathbf{Q}_1^T\mathbf{y} \\ T^2_{\tilde{y}_{LS}} &= \mathbf{y}^T\mathbf{Q}_2\left(\frac{\mathbf{Q}_2^T\mathbf{Y}\mathbf{Y}^T\mathbf{Q}_2}{N-1}\right)^{-1}\mathbf{Q}_2^T\mathbf{y} \end{aligned} \tag{4.11}$$

The corresponding thresholds are

$$\begin{aligned} J_{th,T^2_{\hat{y}_{LS}}} &= \frac{l\left(N^2-1\right)}{N(N-l)}\mathcal{F}_\alpha\left(l, N-l\right) \\ J_{th,T^2_{\tilde{y}_{LS}}} &= \frac{(m-l)\left(N^2-1\right)}{N(N-m+l)}\mathcal{F}_\alpha\left(m-l, N-m+l\right) \end{aligned} \tag{4.12}$$

Then, the decision logic is:

$$\begin{cases} T^2_{\hat{y}_{LS}} \leq J_{th,T^2_{\hat{y}_{LS}}} \text{ and } T^2_{\tilde{y}_{LS}} \leq J_{th,T^2_{\tilde{y}_{LS}}} & \Rightarrow \text{fault} - \text{free} \\ T^2_{\hat{y}_{LS}} > J_{th,T^2_{\hat{y}_{LS}}} & \Rightarrow \text{KPI} - \text{related fault occurs} \\ T^2_{\tilde{y}_{LS}} > J_{th,T^2_{\tilde{y}_{LS}}} & \Rightarrow \text{KPI} - \text{unrelated fault occurs} \end{cases} \tag{4.13}$$

Principal component regression-based method

In [6], a principal component regression (PCR)-based solution for KPI-base PM-FD has been suggested by Ding *et al.* In this approach, PCA is first performed on $\mathbf{Y}$ giving:

$$\mathbf{Y} = \mathbf{P}_{y,pc}\mathbf{T}_{y,pc} + \mathbf{P}_{y,res}\mathbf{T}_{y,res} \tag{4.14}$$

where $\mathbf{P}_{y,pc} \in \mathbb{R}^{m\times\bar{m}}$ and $\mathbf{P}_{y,res} \in \mathbb{R}^{m\times(m-\bar{m})}$ are the loading vectors, $\mathbf{T}_{y,pc}$ denotes the principal score vectors and $\mathbf{T}_{y,res}$ denotes the residual score vectors, $\bar{m}$ is consistent with that defined in Eq. (2.1). $\mathbf{Y}\mathbf{Y}^T$ could be approximately represented with $\mathbf{P}_{y,pc}\boldsymbol{\Lambda}_{y,pc}\mathbf{P}_{y,pc}^T$, where $\boldsymbol{\Lambda}_{y,pc}$ is part of $\boldsymbol{\Lambda}_y$ obtained in Section 3.5.3 and equals $\frac{\mathbf{T}_{y,pc}\mathbf{T}_{y,pc}^T}{N-1}$. The residual part $\mathbf{P}_{y,res}\mathbf{T}_{y,res}$ is automatically determined as the section uncorrelated with KPI. Then the PCR-based regression model can be written as:

$$\hat{\boldsymbol{\theta}}_{PC} = \boldsymbol{\Theta}\mathbf{Y}^T\mathbf{P}_{y,pc}\boldsymbol{\Lambda}_{y,pc}^{-1}\mathbf{P}_{y,pc}^T\mathbf{y} = \bar{\boldsymbol{\Psi}}\bar{\mathbf{y}} \tag{4.15}$$

where $\bar{\mathbf{y}} = \boldsymbol{\Lambda}_{y,pc}^{-1/2}\mathbf{P}_{y,pc}^T\mathbf{y}$, $\bar{\boldsymbol{\Psi}} = \boldsymbol{\Theta}\left(\boldsymbol{\Lambda}_{y,pc}^{-1/2}\mathbf{P}_{y,pc}^T\mathbf{Y}\right)^T$. Subsequently, the following procedures are quite analogous with the LS-based method. Firstly, run a QR decomposition on $\bar{\boldsymbol{\Psi}}^T = \left[\bar{\mathbf{Q}}_1, \bar{\mathbf{Q}}_2\right]\left[\bar{\mathbf{R}}_1^T, 0\right]^T$. Then $\bar{\mathbf{Q}}_1^T\bar{\mathbf{y}}$ satisfying $\bar{\mathbf{Q}}_1^T\bar{y} \sim \mathcal{N}_l\left(0, I_{l\times l}\right)$ is applied to the KPI-related fault detection. The T^2 statistic is

$$T_{\hat{y}_{PC}}^2 = (N-1)\mathbf{y}^T\mathbf{P}_{y,pc}\boldsymbol{\Lambda}_{y,pc}^{-1/2}\bar{\mathbf{Q}}_1\bar{\mathbf{Q}}_1^T\boldsymbol{\Lambda}_{y,pc}^{-1/2}\mathbf{P}_{y,pc}^T\mathbf{y} \tag{4.16}$$

In reality, the KPI-uncorrelated part consists of two separate parts [6]. The first part is the direct residuals of $\mathbf{y}$, namely, $\mathbf{P}_{y,res}\mathbf{P}_{y,res}\mathbf{y}$. It is due to the fact that we have not considered this part to predict the KPI. Another part is identical to the LS-based method, that is $\mathbf{P}_{y,pc}\boldsymbol{\Lambda}_{y,pc}^{1/2}\bar{\mathbf{Q}}_2\bar{\mathbf{Q}}_2^T\boldsymbol{\Lambda}_{y,pc}^{-1/2}\mathbf{P}_{y,pc}\mathbf{y}$. The design of the PM-FD for these two parts is naturally extended along the two different directions. Based on the PM-FD strategy with PCA, the $\mathbf{P}_{y,res}\mathbf{P}_{y,res}\mathbf{y}$ should be monitored using the Q statistic. The second part would be handled by applying the T^2 statistics similarly to LS. Hao *et al.* [9] have extended this idea to the KPI-based PM-FD area, and proposed a combined index for monitoring the KPI-uncorrelated part:

$$T_{\tilde{y}_{PCR}}^2 = \mathbf{y}^T\left((N-1)\lambda_m^2\mathbf{P}_{y,pc}\boldsymbol{\Lambda}_{y,pc}^{-1/2}\bar{\mathbf{Q}}_2\bar{\mathbf{Q}}_2^T\boldsymbol{\Lambda}_{y,pc}^{-1/2}\mathbf{P}_{y,pc}^T + \mathbf{P}_{y,res}\boldsymbol{\Xi}_{res}\mathbf{P}_{y,res}^T\right)\mathbf{y} \tag{4.17}$$

where $\boldsymbol{\Xi}_{res} = \text{diag}\left(\lambda_m^2/\lambda_{\bar{m}+1}^2, ..., \lambda_m^2/\lambda_{m-1}^2, 1\right)$, $\text{diag}\left(\lambda_{\bar{m}+1}^2, ..., \lambda_m^2\right) = \frac{1}{N-1}\mathbf{T}_{res}\mathbf{T}_{res}^T$. The corresponding thresholds for both $T_{\hat{y}_{PCR}}^2$ and $T_{\tilde{y}_{PCR}}^2$ can refer to Eq. (4.7). Finally, the decision logic is

$$\begin{cases} T_{\hat{y}_{PCR}}^2 \leq J_{th,T_{\hat{y}_{PCR}}^2} \text{ and } T_{\tilde{y}_{PCR}}^2 \leq J_{th,T_{\tilde{y}_{PCR}}^2} \Rightarrow \text{fault} - \text{free} \\ T_{\hat{y}_{PCR}}^2 > J_{th,T_{\hat{y}_{PCR}}^2} \Rightarrow \text{KPI} - \text{related fault occurs} \\ T_{\tilde{y}_{PCR}}^2 > J_{th,T_{\tilde{y}_{PCR}}^2} \Rightarrow \text{KPI} - \text{unrelated fault occurs} \end{cases} \tag{4.18}$$

4.2.3 PLS-based methods

In this section, PLS-based solutions are summarized. The methods using original PLS and enhanced PLS (*i.e.*, T-PLS and C-PLS) will be discussed, including the basic ideas and improvements.

Original PLS-based method

The two datasets $\mathbf{Y}$ and $\boldsymbol{\Theta}$ can be related, by PLS, with the score variable $\mathbf{T}$, which are statistically independent, and have a reduced dimension, but they capture the most useful information between $\mathbf{Y}$ and $\boldsymbol{\Theta}$. The PLS algorithm is summarized with two main steps [2]:

- *Step 1*: Set $\mathbf{Y}_1 = \mathbf{Y}$ and recursively calculate for $i = 1, ..., \kappa$

  Step 1.1: $\omega_i^* = \arg \max_{\|\omega_i\|=1} \left\| \mathbf{Y}_i \boldsymbol{\Theta}^T \omega_i \right\|_{\mathrm{E}}^2$

  Step 1.2: $\mathbf{w}_i = \frac{\mathbf{Y}_i \boldsymbol{\Theta}^T \omega_i^*}{\left\| \mathbf{Y}_i \boldsymbol{\Theta}^T \omega_i^* \right\|_{\mathrm{E}}}, \ \mathbf{t}_1 = \mathbf{w}_i^T \mathbf{Y}, \mathbf{p}_i = \mathbf{Y}_i \mathbf{t}_i^T / \mathbf{t}_i \mathbf{t}_i{}^T$

  Step 1.3: $\mathbf{r}_i = \begin{cases} \mathbf{w}_1, i = 1 \\ \prod_{j=1}^{i-1} \left(\mathbf{I}_m - \mathbf{w}_j \mathbf{p}_j^T \right) \mathbf{w}_i, i > 1 \end{cases}, \ \mathbf{q}_i = \boldsymbol{\Theta} \mathbf{t}_i^T / \mathbf{t}_i \mathbf{t}_i{}^T$

  Step 1.4: $\mathbf{Y}_{i+1} = \mathbf{Y}_i - \mathbf{p}_i \mathbf{t}_i$
  where the solution of Step 1.1 is unique and could be achieved by an eigenvalue decomposition [2]. κ is a user-specified known stopping criteria.

- *Step 2*: Form the matrices $\mathbf{T}$, $\mathbf{P}$, $\mathbf{Q}$, and $\mathbf{R}$:
  $\mathbf{T} = [\mathbf{t}_1^T, ..., \mathbf{t}_\kappa^T]^T$, $\mathbf{P} = [\mathbf{p}_1, ..., \mathbf{p}_\kappa]$, $\mathbf{Q} = [\mathbf{q}_1, ..., \mathbf{q}_\kappa]$, $\mathbf{R} = [\mathbf{r}_1, ..., \mathbf{r}_\kappa]$

Let $\mathbf{P}$ denote the loading matrix for $\mathbf{Y}$ and $\mathbf{Q}$ the loading matrix for $\boldsymbol{\Theta}$. $\mathbf{R}$ is the linear transformation, *i.e.*, $\mathbf{T} = \mathbf{R}^T \mathbf{Y}$. After completing the PLS modeling process, $\mathbf{y}$ and $\boldsymbol{\theta}$ are respectively decomposed into $\mathbf{y} = \hat{\mathbf{y}}_{PLS} + \tilde{\mathbf{y}}_{PLS} = \mathbf{P}\mathbf{R}^T \mathbf{y} + \left(\mathbf{I}_m - \mathbf{P}\mathbf{R}^T \right) \mathbf{y}$ and $\boldsymbol{\theta} = \hat{\boldsymbol{\theta}}_{PLS} + \tilde{\boldsymbol{\theta}}_{PLS} = \mathbf{Q}\mathbf{R}^T \mathbf{y} + \tilde{\boldsymbol{\theta}}_{PLS}$.

The classic PM-FD strategy based on PLS applies $\hat{\mathbf{y}}_{PLS}$ to the KPI-related monitoring. Due to the fact that rank$\left(\mathbf{R}^T \hat{\mathbf{Y}}_{PLS} \right) =$

$\text{rank}\left(\mathbf{R}^T\mathbf{Y}\right) = \text{rank}\left(\mathbf{T}\right) = \kappa$, it is reasonable to design a T^2 statistic on $\mathbf{R}^T\mathbf{y}$ in the form of

$$T^2_{\hat{y}_{PLS}} = \mathbf{y}^T\mathbf{R}\left(\frac{\mathbf{T}\mathbf{T}^T}{N-1}\right)^{-1}\mathbf{R}^T\mathbf{y} \tag{4.19}$$

since $\mathbf{R}^T\mathbf{y} \sim \mathcal{N}_\kappa\left(0, \frac{\mathbf{T}\mathbf{T}^T}{N-1}\right)$ holds. The corresponding threshold is based on Eq. (4.12). As for the KPI-uncorrelated part, the Q statistic is recommended for $\tilde{\mathbf{y}}_{PLS}$, especially in the case of $m \gg \kappa$. The index is

$$Q_{PLS} = \tilde{\mathbf{y}}^T_{PLS}\tilde{\mathbf{y}}_{PLS} \tag{4.20}$$

The upper bound is established using Box's theorem [48]: $J_{th,Q_{PLS}} = g\chi^2_{h,\alpha}$, where $g = S/2\mu$, $h = 2\mu/S$. S and μ are calculated according to the method in Chapter 3 [2, 14].

Then, the decision logic is

$$\begin{cases} T^2_{\hat{y}_{PLS}} \le J_{th,T^2_{\hat{y}_{PLS}}} \text{ and } Q \le J_{th,Q} \Rightarrow \text{fault} - \text{free} \\ T^2_{\hat{y}_{PLS}} > J_{th,T^2_{\hat{y}_{PLS}}} \Rightarrow \text{KPI} - \text{related fault occurs} \\ Q > J_{th,Q} \Rightarrow \text{KPI} - \text{unrelated fault occurs} \end{cases} \tag{4.21}$$

T-PLS-based method

As claimed in Remark 3.10, $\tilde{\mathbf{y}}_{PLS}$ still has significant variations, thus, is not well-suited to apply Q-statistic. Furthermore, Zhou *et al.* [37] proved that there are KPI-uncorrelated parts in $\hat{\mathbf{y}}_{PLS}$. T-PLS solves the two problem by further decomposing $\hat{\mathbf{Y}}_{PLS}$ and $\tilde{\mathbf{Y}}_{PLS}$. The step-wise T-PLS algorithm is briefly summarised in the following steps:

- *Step 1*: Run the PLS algorithm to obtain $\hat{\mathbf{Y}}_{PLS}$, $\tilde{\mathbf{Y}}_{PLS}$ and $\hat{\mathbf{\Theta}}_{PLS}$
- *Step 2*: Divide $\hat{\mathbf{Y}}_{PLS}$ supervised by $\hat{\mathbf{\Theta}}_{PLS}$:

  Step 2.1: Perform a PCA on $\hat{\mathbf{\Theta}}_{PLS}$: $\hat{\mathbf{\Theta}}_{PLS} = \mathbf{Q}_\theta\mathbf{T}_\theta$ and $\mathbf{T}_\theta = \mathbf{Q}^T_\theta\hat{\mathbf{\Theta}}_{PLS}$

  Step 2.2: Formulate $\mathbf{Y}_\theta$ using $\mathbf{T}_\theta$ and $\hat{\mathbf{Y}}_{PLS}$:
  $$\mathbf{Y}_\theta = \mathbf{P}_\theta\mathbf{T}_\theta = \hat{\mathbf{Y}}_{PLS}\mathbf{T}^T_\theta\left(\mathbf{T}_\theta\mathbf{T}^T_\theta\right)^{-1}\mathbf{T}_\theta$$

  Step 2.3: Perform PCA on $\mathbf{Y}_o = \hat{\mathbf{Y}}_{PLS} - \mathbf{Y}_\theta$, then $\mathbf{Y}_o = \mathbf{P}_o\mathbf{T}_o$ and $\mathbf{T}_o = \mathbf{P}^T_o\mathbf{Y}_o$.

- *Step 3*: Further model $\tilde{\mathbf{Y}}_{PLS}$ with PCA yielding
$\tilde{\mathbf{Y}}_{PLS} = \mathbf{Y}_r + \mathbf{Y}_{rr} = \mathbf{P}_r\mathbf{T}_r + \mathbf{Y}_{rr}\left(= \mathbf{P}_{rr}\mathbf{T}_{rr}\right), \ \mathbf{T}_r = \mathbf{P}_r^T\tilde{\mathbf{Y}}_{PLS}$

In Step 2.1, the PCA keeps rank($\mathbf{Q}$) (in general, it is equivalent to l) principal components of $\hat{\mathbf{\Theta}}_{PLS}$, Step 2.3 reserves κ − rank($\mathbf{Q}$) principal components of $\mathbf{Y}_o$. In Step 3, the principal component number κ_r is set beforehand (see [37] for details).

To be analogous with PLS, the score vectors $\mathbf{t}_\theta$, $\mathbf{t}_o$ and $\mathbf{t}_r$ play the main role for designing PM-FD methods. Let denote $\mathbf{R}_\theta$, $\mathbf{R}_o$ and $\mathbf{R}_r$ as the direct weighting matrices such that $\mathbf{R}_\theta^T\mathbf{y}, \mathbf{R}_o^T\mathbf{y}, \mathbf{R}_r^T\mathbf{y}$ [37]. The monitoring indices based on T-PLS are comprehensive. $\mathbf{Y}_\theta$ is represented using $\mathbf{t}_\theta$, so it is able to build a T_θ^2 statistic on $\mathbf{t}_\theta$: $T_\theta^2 = \mathbf{t}_\theta^T\left(\frac{\mathbf{T}_\theta\mathbf{T}_\theta^T}{N-1}\right)^{-1}\mathbf{t}_\theta$. $\mathbf{Y}_o$ could be monitored based on the T^2 statistic on $\mathbf{t}_o$, namely $T_o^2 = \mathbf{t}_o^T\left(\frac{\mathbf{T}_o\mathbf{T}_o^T}{N-1}\right)^{-1}\mathbf{t}_o$. T_r^2 is similarly obtained with $T_r^2 = \mathbf{t}_r^T\left(\frac{\mathbf{T}_r\mathbf{T}_r^T}{N-1}\right)^{-1}\mathbf{t}_r$. $\mathbf{y}_{rr}$ contains small variations that should be monitored by the Q_{T-PLS} statistic.

The decision logic is

$$\begin{cases} T_\theta^2 \le J_{th,T_\theta^2} \text{ and } T_o^2 \le J_{th,T_o^2} \text{ and } T_r^2 \le J_{th,T_r^2} \text{ and } Q_{T-PLS} \le J_{th,Q_{T-PLS}} \Rightarrow \text{fault}-\text{free} \\ T_\theta^2 > J_{th,T_\theta^2} \text{ or } Q_{T-PLS} > J_{th,Q_{T-PLS}} \Rightarrow \text{KPI}-\text{related fault occurs} \\ T_o^2 > J_{th,T_o^2} \text{ or } T > J_{th,T_r^2} \Rightarrow \text{KPI}-\text{unrelated fault occurs} \end{cases} \tag{4.22}$$

C-PLS-based method

C-PLS is a new approach which attempts to solve the KPI-based PM-FD issue more accurately and efficiently [24]. The C-PLS algorithm consists of the following steps:

- *Step 1*: Build the PLS model on $\mathbf{Y}$ and $\mathbf{\Theta}$ to obtain $\hat{\mathbf{\Theta}}_{PLS}$
- *Step 2*: Remodel $\mathbf{Y}$ supervised by $\hat{\mathbf{\Theta}}_{PLS}$: $\mathbf{Y} = \mathbf{Y}_{\hat{\theta}} + \mathbf{Y}_{\hat{\theta}^\perp}$:

  Step 2.1: Build PCA model on $\hat{\mathbf{\Theta}}_{PLS}$ to extract the score $\mathbf{T}_\theta$: $\mathbf{T}_\theta = \mathbf{R}_\theta^T Y$.

  Step 2.2: Project $\mathbf{Y}$ onto the orthogonal subspaces to obtain $\mathbf{Y}_{\hat{\theta}}$ and $\mathbf{Y}_{\hat{\theta}^\perp}$:
  $\mathbf{Y}_{\hat{\theta}} = (\mathbf{R}_\theta^T)^\dagger\mathbf{R}_\theta^T\mathbf{Y}$ and $\mathbf{Y}_{\hat{\theta}^\perp} = \left(\mathbf{I}_m - (\mathbf{R}_\theta^T)^\dagger\mathbf{R}_\theta^T\right)\mathbf{Y}$

- *Step 3*: Perform a PCA on $\mathbf{Y}_{\hat{\theta}\perp}$: $\mathbf{Y}_{\hat{\theta}\perp} = \mathbf{Y}_{\theta\perp} + \mathbf{Y}_{\tilde{\theta}}$ where $\hat{\mathbf{Y}}_{\theta\perp}$ could be interpreted by $\mathbf{T}_{\theta\perp}$, $\mathbf{T}_{\theta\perp} = \mathbf{R}_{\theta\perp}^T\mathbf{Y}$, $\mathbf{Y}_{\theta\perp} = \mathbf{P}_{\theta\perp}\mathbf{T}_{\theta\perp}$.

In Step 2.1, $\mathbf{R}_\theta$ is consistent with the one in T-PLS. In Step 3, the PCA contains $\kappa_{\theta\perp}$ PCs, which is determined by the PCA-based methods [24]. $\mathbf{Y}_{\tilde{\theta}}$ is spanned by $\mathbf{P}_{\tilde{\theta}}$

T^2 statistic can be used with $Y_{\hat{\theta}}$, namely, $T_\theta^2 = \mathbf{t}_\theta^T\left(\frac{\mathbf{T}_\theta\mathbf{T}_\theta^T}{N-1}\right)^{-1}\mathbf{t}_\theta$. Q statistic is more suitable for $\mathbf{Y}_{\tilde{\theta}}$, *i.e.*, $Q_{\tilde{y}_\theta} = \tilde{\mathbf{y}}_\theta^T\tilde{\mathbf{y}}_\theta$ could be set. For $\mathbf{Y}_{\theta\perp}$, which is produced by a PCA model, therefore a suitable T^2 statistic: $T_{y_{\theta\perp}}^2 = \mathbf{t}_{\theta\perp}^T\left(\frac{\mathbf{T}_{\theta\perp}\mathbf{T}_{\theta\perp}^T}{N-1}\right)^{-1}\mathbf{t}_{\theta\perp}$ is given. All of their thresholds are obtained by referring to the aforementioned approaches.

Thus, the decision logic is given by

$$\begin{cases} T_\theta^2 \leq J_{th,T_\theta^2} \text{ and } T_{\theta\perp}^2 \leq J_{th,T_{\theta\perp}^2} \text{ and } Q_{C-PLS} \leq J_{th,Q_{C-PLS}} \Rightarrow \text{fault}-\text{free} \\ T_\theta^2 > J_{th,T_\theta^2} \text{ or } Q_{T-PLS} > J_{th,Q_{T-PLS}} \Rightarrow \text{KPI}-\text{related fault occurs} \\ T_{\theta\perp}^2 > J_{th,T_{\theta\perp}^2} \Rightarrow \text{KPI}-\text{unrelated fault occurs} \end{cases} \tag{4.23}$$

4.3 Theoretical comparisons

4.3.1 Interconnections among the approaches

Similar to PCA, the direct method can be considered as a direct SVD solution for the PM-FD problem with KPIs involved. In this method, the following cross-covariance formulas are satisfied: $\mathrm{E}\left(\tilde{\mathbf{y}}\boldsymbol{\theta}^T\right) = 0$, and $\mathrm{E}\left(\hat{\mathbf{y}}\boldsymbol{\theta}^T\right) = \mathrm{E}\left(\mathbf{y}\boldsymbol{\theta}^T\right)$. In LS-based approach, $\boldsymbol{\Psi}_{LS}\hat{\mathbf{y}}_{LS} = \boldsymbol{\Psi}_{LS}\mathbf{y}$ and $\boldsymbol{\Psi}_{LS}\tilde{\mathbf{y}}_{LS} = 0$, and it should be noted that $T_{\hat{\theta}_{LS}}^2 = \mathbf{y}^T\boldsymbol{\Psi}_{LS}^T\left(\frac{\boldsymbol{\Psi}_{LS}\mathbf{Y}\mathbf{Y}^T\boldsymbol{\Psi}_{LS}^T}{N-1}\right)^{-1}\boldsymbol{\Psi}_{LS}\mathbf{y} = \mathbf{y}^T\mathbf{Q}_1\mathbf{R}_1\left(\frac{\mathbf{R}_1^T\mathbf{Q}_1^T\mathbf{Y}\mathbf{Y}^T\mathbf{Q}_1\mathbf{R}_1}{N-1}\right)^{-1}\mathbf{R}_1^T\mathbf{Q}_1^T\mathbf{y} = T_{\hat{y}_{LS}}^2$, where $T_{\hat{\theta}_{LS}}^2 = \boldsymbol{\theta}_{LS}^T\left(\frac{\hat{\boldsymbol{\Theta}}_{LS}\hat{\boldsymbol{\Theta}}_{LS}^T}{N-1}\right)^{-1}\hat{\boldsymbol{\theta}}_{LS}$. This implies that the design based on LS keeps consistency with the principle of KPI-based PM-FD, namely, when there is no online KPI available, the predicted KPI (*i.e.*, $\hat{\boldsymbol{\theta}}_{LS}$) can be applied instead. Similar to LS, $T^2_{\hat{y}_{PC}} = T^2_{\hat{\theta}_{pc}} \triangleq \hat{\boldsymbol{\theta}}^T_{pc}\left(\frac{\hat{\boldsymbol{\Theta}}_{pc}\hat{\boldsymbol{\Theta}}^T_{pc}}{N-1}\right)^{-1}\hat{\boldsymbol{\theta}}_{pc}$. This point links the KPI-based PM-FD to linear regression methods, namely, aimed at real-time monitoring KPI, the results achieved by them, as well, obey the law that the partitioned KPI-correlated part serves for predicting KPI. The differences are that LS derives the $\hat{\mathbf{y}}_{LS}$ based on the LS regression matrix $\boldsymbol{\Psi}_{LS}$, which can be directly identified from the data sets $\mathbf{Y}$ and $\boldsymbol{\Theta}$, while PCR based method seeks to find $\hat{\mathbf{y}}_{PCR}$ indirectly from the principal space of $\mathbf{y}$ ($\mathbf{P}_{y,pc}$) obtained by a PCA decomposition on $\mathbf{Y}$. This difference leads to the fact that $\hat{\mathbf{y}}_{PC}$ and $\tilde{\mathbf{y}}_{PC}$ are not geometrically orthogonal, but are statistically independent since $\mathrm{E}\left(\hat{\mathbf{y}}_{pc}\tilde{\mathbf{y}}^T_{pc}\right) = 0$. LS-based method involves more statistical characteristics than PCR. It can be seen from Eq. (4.9) that two covariance estimates: $\mathrm{E}\left(\boldsymbol{\theta}\mathbf{y}^T\right)$ and $\mathrm{E}\left(\mathbf{y}\mathbf{y}^T\right)$ are solely required for LS-based method, while in Eq. (4.15), there is no such term for PCR. Thus, for the process with multiple sampling rate for $\mathbf{Y}$ and $\boldsymbol{\Theta}$, LS can also work by using part of $\mathbf{Y}$ and $\boldsymbol{\Theta}$ to estimate $\mathrm{E}\left(\boldsymbol{\theta}\mathbf{y}^T\right)$, and all $\mathbf{Y}$ to estimate $\mathrm{E}\left(\mathbf{y}\mathbf{y}^T\right)$, but PCR cannot. The PCA decomposition in PCR makes it strongly against the overfitting, a common problem usually occurs in LS. It further makes the detection statistics of PCR less sensitive to process noises. Thus, compared with LS, PCR-based method is more robust to system noises. Regarding the linear regression-based methods, it should be noted that some other linear regression approaches (*e.g.*, ridge regression [56], canonical correlation regression [57, 82]) can be introduced into this class in the similar way.

The statistical properties could be particularly obtained in PLS model:

$$\begin{aligned}
\mathrm{E}\left(\mathbf{y}\mathbf{y}^T\right) &= \mathrm{E}\left(\hat{\mathbf{y}}_{PLS}\hat{\mathbf{y}}^T_{PLS}\right) + \mathrm{E}\left(\tilde{\mathbf{y}}_{PLS}\tilde{\mathbf{y}}^T_{PLS}\right)\\
\mathrm{E}\left(\boldsymbol{\theta}\boldsymbol{\theta}^T\right) &= \mathrm{E}\left(\hat{\boldsymbol{\theta}}_{PLS}\hat{\boldsymbol{\theta}}^T_{PLS}\right) + \mathrm{E}\left(\tilde{\boldsymbol{\theta}}_{PLS}\tilde{\boldsymbol{\theta}}^T_{PLS}\right)\\
\mathrm{E}\left(\hat{\mathbf{y}}_{PLS}\tilde{\boldsymbol{\theta}}^T_{PLS}\right) &= 0, \mathrm{E}\left(\tilde{\mathbf{y}}_{PLS}\hat{\boldsymbol{\theta}}^T_{PLS}\right) = 0
\end{aligned} \tag{4.24}$$

Further notice that $\mathrm{E}\left(\tilde{\mathbf{y}}_{PLS}\tilde{\boldsymbol{\theta}}^T_{PLS}\right) \approx 0$ or $\mathrm{E}\left(\tilde{\mathbf{y}}_{PLS}\tilde{\boldsymbol{\theta}}^T_{PLS}\right) \neq 0$, which implies there are potential correlations between them. For PLS-based PM-FD, Li *et al.* have shown that $T^2_{\hat{y}_{PLS}} = \hat{\mathbf{y}}^T_{PLS}\left(\frac{\hat{\mathbf{Y}}_{PLS}\hat{\mathbf{Y}}^T_{PLS}}{N-1}\right)^{\dagger}\hat{\mathbf{y}}_{PLS}$ [29]. However, it is observed that $T^2_{\hat{y}_{PLS}} \neq \mathbf{y}^T\mathbf{R}\mathbf{Q}^T\left(\frac{\mathbf{Q}\mathbf{R}^T\mathbf{Y}\mathbf{Y}^T\mathbf{R}\mathbf{Q}^T}{N-1}\right)^{-1}\mathbf{Q}\mathbf{R}^T\mathbf{y} = \boldsymbol{\theta}^T_{PLS}\left(\frac{\boldsymbol{\Theta}_{PLS}\boldsymbol{\Theta}^T_{PLS}}{N-1}\right)^{-1}\boldsymbol{\theta}_{PLS}$. This point indicates that unlike in the LS-based approach, PM-FD with PLS does not use all the available information obtained from PLS, hence, it is not efficient. For more discussion on this topic, one can refer to [2]. The motivation of T-PLS lies in the disadvantage of PLS when applied to FM-FD. On the one hand, $\hat{\mathbf{y}}_{PLS}$ includes redundant information uncorrelated to $\hat{\boldsymbol{\theta}}_{PLS}$, while on the other hand, as mentioned before, $\tilde{\mathbf{y}}_{PLS}$ is weakly correlated with $\tilde{\boldsymbol{\theta}}_{PLS}$, which may affect $\mathbf{y}$ as well. Note that in T-PLS, $T_\theta^2 = T^2_{\hat{\theta}_{PLS}} \triangleq \hat{\boldsymbol{\theta}}_{PLS}^T \left(\frac{\hat{\boldsymbol{\Theta}}_{PLS}\hat{\boldsymbol{\Theta}}_{PLS}^T}{N-1}\right)^{-1} \hat{\boldsymbol{\theta}}_{PLS}$, which is the same as LS-regression based approach. C-PLS, on the one hand, resembles the LS-based method to obtain the direct KPI-correlated part. On the other hand, it inherits the spirit of T-PLS to consider the likely KPI-correlated part. To be more convenient, compared with the four subspaces from T-PLS, it merely decomposes $\mathbf{y}$ into three subspaces. Regarding the PM-FD based on C-PLS, T_θ^2 statistic is identical with its counterpart in T-PLS. Also, note that $\mathbf{Y}_{\hat{\theta}} = \mathbf{R}_\theta(\mathbf{R}_\theta^T\mathbf{R}_\theta)^{-1}\mathbf{R}_\theta^T\mathbf{Y} = \mathbf{R}_\theta\mathbf{Q}_\theta^T(\mathbf{Q}_\theta\mathbf{R}_\theta^T\mathbf{R}_\theta\mathbf{Q}_\theta^T)^{-1}\mathbf{Q}_\theta\mathbf{R}_\theta^T\mathbf{Y} = \mathbf{\Pi}_{\mathbf{\Psi}^T_{\mathbf{PLS}}}Y$, which fairly resembles the property in LS-based approach, that is to say, if $\mathbf{\Psi}_{PLS} = \mathbf{\Psi}_{LS}$, $\mathbf{Y}_{\hat{\boldsymbol{\theta}}} = \hat{\mathbf{Y}}_{LS}$ holds. Based on [2, 29] and Lemma 3 in [37], there exists a linear relationship $\mathbb{M} = \begin{bmatrix} \mathbf{Q}_\theta\mathbf{Q} \\ \mathbf{P}_o^T\left(\mathbf{P} - \mathbf{P}_\theta\mathbf{Q}_\theta\mathbf{Q}\right) \end{bmatrix} \in \mathbb{R}^{\kappa\times\kappa}$ such that $\begin{bmatrix} \mathbf{t}_\theta \\ \mathbf{t}_o \end{bmatrix} = \mathbb{M}\mathbf{t}_{PLS}$, where $\mathbf{t}_{PLS} = \mathbf{R}^T\mathbf{y}$ is the score of the PLS model. This implies that the T^2 statistic in the PLS-based method can be rewritten as

$$\begin{aligned} T^2_{\hat{y}_{PLS}} &= \mathbf{t}_{PLS}^T\Sigma^{-1}_{\mathbf{t}_{PLS}}\mathbf{t}_{PLS} \\ &= \mathbf{t}_{PLS}^T\mathbb{M}^T\left(\mathbb{M}\Sigma_{\mathbf{t}_{PLS}}\mathbb{M}^T\right)^{-1}\mathbb{M}\mathbf{t}_{PLS} \\ &= \mathbf{t}_\theta^T\Sigma^{-1}_{\mathbf{t}_\theta}\mathbf{t}_\theta + \mathbf{t}_o^T\Sigma^{-1}_{\mathbf{t}_o}\mathbf{t}_o \\ &= \sum_{i=1}^{l}\frac{\mathbf{t}^2_{\theta,i}}{\sigma^2_{\theta,i}} + \sum_{i=l+1}^{\kappa}\frac{\mathbf{t}^2_{o,i}}{\sigma^2_{o,i}} \\ &= T_\theta^2 + T_o^2 \\ &\sim \chi^2_\kappa \end{aligned} \tag{4.25}$$

where $\Sigma_{\mathbf{t}_{PLS}} = \frac{\mathbf{T}\mathbf{T}^T}{N-1}$ as shown in Eq. (4.19), $\sigma^2_{\theta,i}$ and $\sigma^2_{o,i}$ are similar to that in T-PLS method. As shown in T-PLS and C-PLS methods, $T_\theta^2 = T^2_{\hat{\theta}_{PLS}} = T^2_{\hat{\theta}_{LS}}$. In the case of an additive fault that can directly affect the KPI, this part will be straightforwardly influenced with $\hat{\boldsymbol{\theta}}_{LS,f} = \hat{\boldsymbol{\theta}}_{PLS,f}$. Using $T^2_{\hat{y}_{PLS}}$ and $T^2_{\hat{y}_{LS}}$, we can see that the fault gives the same noncentrality parameter in Eq. (3.9), namely $\delta_{LS} = \delta_{PLS}$. It, thus, leads to

$$\mathrm{FDR}_{LS} = \mathrm{prob}\left(\chi_l^2(\delta_{LS}) > \chi_{l,\alpha}^2\right) \geq \mathrm{prob}\left(\chi_\kappa^2(\delta_{PLS}) > \chi_{\kappa,\alpha}^2\right) = \mathrm{FDR}_{PLS} \tag{4.26}$$

with $\kappa \geq l$. If the PLS model does not completely capture the KPI-relevant models, *i.e.* $\mathbf{\Psi}_{PLS} \neq \mathbf{\Psi}_{LS}$, then there exists the uncovered part by PLS which can still affect KPI. If a fault occurs in this part, PLS-based method will not detect it. In contrast, LS can capture this part and detect such faults. Note that C-PLS and T-PLS were proposed aiming at solving this specific problem. However, both of them cannot determine whether PLS has captured most of the KPI-correlated information in $\mathbf{y}$ or not, and blindly apply the noisy part to be a potential for monitoring the KPI. It is certain that a large amount of false alarms will be produced by them. Furthermore, the two methods perform a simple principal component analysis on the KPI-irrelevant part of the PLS, and the part with minor variances are assumed to include the KPI information uncovered by PLS. However, PCA can only isolate the principal part based on the variance in the components, thus it is likely that the uncovered part may be assigned to the part with major variances.

4.3.2 Geometric properties and computations

Geometric properties

In the direct method, define two orthogonal projections $\hat{\mathbf{y}} = \mathbf{U}_{\theta y}\mathbf{U}_{\theta y}^T\mathbf{y}$ and $\tilde{\mathbf{y}} = \mathbf{U}_{\theta y^\perp}\mathbf{U}_{\theta y^\perp}^T\mathbf{y}$, then $\mathbf{\Pi}_\theta = \mathbf{U}_{\theta y}\mathbf{U}_{\theta y}^T$ and $\mathbf{\Pi}_{\theta^\perp} = \mathbf{U}_{\theta y^\perp}\mathbf{U}_{\theta y^\perp}^T$. Meanwhile $\mathbb{S}_\theta = \mathrm{span}\{\mathbf{U}_{\theta y}\}$ and $\mathbb{S}_{\theta^\perp} = \mathrm{span}\{\mathbf{U}_{\theta y^\perp}\}$. $\hat{\mathbf{y}}$ covers the $\boldsymbol{\theta}$-correlated part in $\mathbf{y}$, while $\tilde{\mathbf{y}}$ belongs to the orthogonal supplement space of $\hat{\mathbf{y}}$ and completely uncorrelated to $\boldsymbol{\theta}$. It can be seen in LS-based method that $\mathrm{span}\{\mathbf{\Psi}_{PLS}^T\} = \mathrm{span}\{\mathbf{Q}_1\}$ and $\mathrm{span}\{\mathbf{\Psi}_{PLS}^T\}^\perp = \mathrm{span}\{\mathbf{Q}_2\}$. The two projections are set as $\hat{\mathbf{y}}_{LS} = \mathbf{Q}_1\mathbf{Q}_1^T\mathbf{y}$ and $\tilde{\mathbf{y}}_{LS} = \mathbf{Q}_2\mathbf{Q}_2^T\mathbf{y}$ with $\mathbf{\Pi}_\theta = \mathbf{Q}_1\mathbf{Q}_1^T$ and $\mathbf{\Pi}_{\theta^\perp} = \mathbf{Q}_2\mathbf{Q}_2^T$. Thus $\mathbb{S}_\theta = \mathrm{span}\{\mathbf{Q}_1\}$ and $\mathbb{S}_{\theta^\perp} = \mathrm{span}\{\mathbf{Q}_2\}$. Define $\mathbf{P}_{pc} = \mathbf{P}_{y,pc}\mathbf{\Lambda}_{y,pc}^{1/2}\bar{\mathbf{Q}}_1$ and $\mathbf{R}_{pc} = \mathbf{P}_{y,pc}\mathbf{\Lambda}_{y,pc}^{-1/2}\bar{\mathbf{Q}}_1$. Note that $\mathbf{R}_{pc}^T\mathbf{P}_{pc} = \mathbf{P}_{pc}^T\mathbf{R}_{pc} = \mathbf{I}_l$ holds. The following theorem shows the setting for the projection structure induced by the PCR-based approach.

Theorem 4.1. *The projection along* $\text{span}\{\mathrm{R}_{pc}\}^{\perp}$ *onto* $\text{span}\{\mathbf{P}_{pc}\}$ *is written as* $\mathbf{P}_{pc}\mathbf{R}_{pc}{}^{T}$. *Let* $\mathbf{\Pi}_{\theta} = \mathbf{P}_{pc}\mathbf{R}_{pc}{}^{T}$ *and* $\hat{\mathbf{y}}_{PC} = \mathbf{\Pi}_{\theta}\mathbf{y}$, *then* $T^2_{\hat{y}_{PC}} = \hat{\mathbf{y}}_{PC}^{T}\left(\frac{\hat{\mathbf{Y}}_{PC}\hat{\mathbf{Y}}_{PC}^{T}}{N-1}\right)^{\dagger}\hat{\mathbf{y}}_{PC}$ *holds.*

Proof. The theorem can be similarly proved using the method mentioned in [29]. □

Note that the KPI-correlated projector induced by PCR resembles the one in the PLS model, which will be discussed below. In addition, along this direction, the KPI-uncorrelated part is of course obtained though $\tilde{\mathbf{y}}_{PC} = \left(\mathbf{I}_m - \mathbf{P}_{pc}\mathbf{R}_{pc}^{T}\right)\mathbf{y}$, namely, $\mathbf{\Pi}_{\theta^{\perp}} = \mathbf{I}_m - \mathbf{P}_{pc}\mathbf{R}_{pc}^{T}$. Then, $S_{\theta^{\perp}}$ is equal to $\text{span}\{\mathbf{R}_{pc}\}^{\perp}$. It is also reasonable to conclude that $\text{span}\{\mathbf{R}_{pc}\}^{\perp} = \text{span}\left\{\mathbf{P}_{y,pc}\mathbf{\Lambda}_{y,pc}^{1/2}\bar{\mathbf{Q}}_2\right\} \oplus \text{span}\{\mathbf{P}_{res}\}$. Li *et al.* [29] claim, for PLS model, that $\mathbf{\Pi}_{\theta} = \mathbf{P}\mathbf{R}^{T}$, $\mathbf{\Pi}_{\theta^{\perp}} = \mathbf{I}_m - \mathbf{P}\mathbf{R}^{T}$, and $\mathbb{S}_{\theta} = \text{span}\{\mathbf{P}\}$, $\mathbb{S}_{\theta^{\perp}} = \text{span}\{\mathbf{R}\}^{\perp}$. $\hat{\theta}_{PLS}$ stands for the predictable part of θ with PLS. T-PLS decomposes $\mathbf{y}$ into four subsections, *i.e.* $\mathbf{y}_{\theta}$, $\mathbf{y}_{o}$, $\mathbf{y}_{r}$ and $\mathbf{y}_{rr}$, of which $\mathbf{y}_{\theta}$ and $\mathbf{y}_{rr}$ are considered as parts correlated to KPI, *i.e.* $\hat{\mathbf{y}}_{T-PLS} = \mathbf{y}_{\theta} + \mathbf{y}_{rr}$, $\mathbf{y}_{o}$ and $\mathbf{y}_{r}$ are the KPI-uncorrelated parts, *i.e.* $\tilde{\mathbf{y}}_{T-PLS} = \mathbf{y}_{o} + \mathbf{y}_{r}$. The four corresponding projectors are

$$\begin{cases} \mathbf{\Pi}_{\theta,1} = \mathbf{P}_{\theta}\mathbf{R}_{\theta}^{T} \\ \mathbf{\Pi}_{\theta^{\perp},1} = \mathbf{P}\mathbf{R}^{T} - \mathbf{P}_{\theta}\mathbf{R}_{\theta}^{T} \\ \mathbf{\Pi}_{\theta^{\perp},2} = \mathbf{P}_{r}\mathbf{R}_{r}^{T} \\ \mathbf{\Pi}_{\theta,2} = \mathbf{I}_m - \mathbf{P}\mathbf{R}^{T} - \mathbf{P}_{r}\mathbf{R}_{r}^{T} \end{cases} \tag{4.27}$$

The following theorem further explains the four subspaces.

Theorem 4.2. $\mathbf{P}_{\theta}\mathbf{R}_{\theta}^{T}$ *induces a projector along the* $\text{span}\{\mathbf{R}_{\theta}\}^{\perp}$ *onto* $\text{span}\{\mathbf{P}_{\theta}\}$, *while* $\mathbf{P}_{r}\mathbf{R}_{r}^{T}$ *induces a projector long* $\text{span}\{\mathbf{R}_{r}\}^{\perp}$ *onto* $\text{span}\{\mathbf{P}_{r}\}$. $\mathbf{P}\mathbf{R}^{T} - \mathbf{P}_{\theta}\mathbf{R}_{\theta}^{T}$ *and* $\mathbf{I}_m - \mathbf{P}\mathbf{R}^{T} - \mathbf{P}_{r}\mathbf{R}_{r}^{T}$ *are the projectors projecting* $\mathbf{y}$ *onto their relevant complement spaces.*

Proof. The oblique projector onto $\text{span}\{\mathbf{P}_{\theta}\}$ along $\text{span}\{\mathbf{R}_{\theta}\}^{\perp}$ is

$$\begin{aligned} \mathbf{\Pi}_{\mathbf{P}_{\theta}|\mathbf{R}_{\theta}^{\perp}} &= \mathbf{P}_{\theta}\left(\mathbf{P}_{\theta}^{T}\mathbf{\Pi}_{\mathbf{R}_{\theta}^{\perp}}^{\perp}\mathbf{P}_{\theta}\right)^{-1}\mathbf{P}_{\theta}{}^{T}\mathbf{\Pi}_{\mathbf{R}_{\theta}^{\perp}}^{\perp} \\ &= \mathbf{P}_{\theta}\left(\mathbf{P}_{\theta}^{T}\mathbf{R}_{\theta}\left(\mathbf{R}_{\theta}^{T}\mathbf{R}_{\theta}\right)^{-1}\mathbf{R}_{\theta}^{T}\mathbf{P}_{\theta}\right)^{-1}\mathbf{P}_{\theta}^{T}\mathbf{R}_{\theta}\left(\mathbf{R}_{\theta}^{T}\mathbf{R}_{\theta}\right)^{-1}\mathbf{R}_{\theta}^{T} \end{aligned} \tag{4.28}$$

Note that $\mathbf{P}_\theta^T\mathbf{R}_\theta = \mathbf{P}_\theta^T\mathbf{R}\mathbf{Q}^T\mathbf{Q}_\theta = \left(\mathbf{T}_\theta\mathbf{T}_\theta^T\right)^{-1}\mathbf{T}_\theta\hat{\mathbf{Y}}_{PLS}^T\mathbf{R}\mathbf{Q}^T\mathbf{Q}_\theta = \left(\mathbf{T}_\theta\mathbf{T}_\theta^T\right)^{-1}\mathbf{T}_\theta\mathbf{T}_\theta^T = \mathbf{I}_l$. Thus, $\mathbf{\Pi}_{\mathbf{P}_\theta|\mathbf{R}_\theta^\perp} = \mathbf{P}_\theta\mathbf{R}_\theta^T$.

Quite similarly, the oblique projector onto span$\{\mathbf{P}_r\}$ along span$\{\mathbf{R}_r\}^\perp$ is derived as

$$\begin{aligned}\mathbf{\Pi}_{\mathbf{P_r}|\mathbf{R_r^\perp}} &= \mathbf{P}_r\left(\mathbf{P}_r^T\mathbf{\Pi}_{\mathbf{R_r^\perp}}^\perp\mathbf{P_r}\right)^{-1}\mathbf{P}_r^T\mathbf{\Pi}_{\mathbf{R_r^\perp}}^\perp \\ &= \mathbf{P}_r\left(\mathbf{P}_r^T\mathbf{R}_r\left(\mathbf{R}_r^T\mathbf{R}_r\right)^{-1}\mathbf{R}_r^T\mathbf{P}_r\right)^{-1}\mathbf{P}_r^T\mathbf{R}_r\left(\mathbf{R}_r^T\mathbf{R}_r\right)^{-1}\mathbf{R}_r^T\end{aligned} \tag{4.29}$$

Meanwhile, $\mathbf{P}_r^T\mathbf{R}_r = \mathbf{P}_r^T\left(\mathbf{I}_m - \mathbf{R}\mathbf{P}^T\right)\mathbf{P}_r = \mathbf{I}_m - \mathbf{P}_r^T\mathbf{R}\mathbf{P}^T\mathbf{P}_r$ holds. As well from the T-PLS model, $\mathbf{P}_r \in$ span$\{\mathbf{R}\}^\perp$ is satisfied. Hence $\mathbf{P}_r^T\mathbf{R} = 0$ holds automatically, which leads to the fact that $\mathbf{P}_r^T\mathbf{R}_r = \mathbf{I}_{\kappa_r}$, so $\mathbf{\Pi}_{\mathbf{P_r}|\mathbf{R_r^\perp}} = \mathbf{P}_r\mathbf{R}_r^T$ is proven.

Additionally, $\left(\mathbf{P}\mathbf{R}^T - \mathbf{P}_\theta\mathbf{R}_\theta^T\right)\left(\mathbf{P}\mathbf{R}^T - \mathbf{P}_\theta\mathbf{R}_\theta^T\right) = \mathbf{P}\mathbf{R}^T\mathbf{P}\mathbf{R}^T - \mathbf{P}_\theta\mathbf{R}_\theta^T\mathbf{P}\mathbf{R}^T - \mathbf{P}\mathbf{R}^T\mathbf{P}_\theta\mathbf{R}_\theta^T + \mathbf{P}_\theta\mathbf{R}_\theta^T\mathbf{P}_\theta\mathbf{R}_\theta^T$, where $\mathbf{R}^T\mathbf{P} = \mathbf{I}_m$ comes from PLS model, and

$$\begin{aligned}\mathbf{P}\mathbf{R}^T\mathbf{P}_\theta\mathbf{R}_\theta^T &= \mathbf{P}\mathbf{R}^T\hat{\mathbf{Y}}_{PLS}\mathbf{T}_\theta^T\left(\mathbf{T}_\theta\mathbf{T}_\theta^T\right)^{-1}\mathbf{R}_\theta^T = \mathbf{P}_\theta\mathbf{R}_\theta^T \\ \mathbf{P}_\theta\mathbf{R}_\theta^T\mathbf{P}\mathbf{R}^T &= \mathbf{P}_\theta\mathbf{Q}_\theta^T\mathbf{Q}\mathbf{R}^T\mathbf{P}\mathbf{R}^T = \mathbf{P}_\theta\mathbf{R}_\theta^T\end{aligned}$$

Therefore, $\mathbf{P}\mathbf{R}^T - \mathbf{P}_\theta\mathbf{R}_\theta^T$ is idempotent and naturally acts as a projector. Furthermore, the two projectors satisfy $\left(\mathbf{P}\mathbf{R}^T - \mathbf{P}_\theta\mathbf{R}_\theta^T\right)\mathbf{P}_\theta\mathbf{R}_\theta^T = 0$, which shows that they project $\mathbf{y}$ onto two overlapped subspaces. As $\left(\mathbf{P}\mathbf{R}^T - \mathbf{P}_\theta\mathbf{R}_\theta^T\right)\mathbf{y} + \mathbf{P}_\theta\mathbf{R}_\theta^T\mathbf{y} = \mathbf{P}\mathbf{R}^T\mathbf{y}$, and $\mathbf{P}\mathbf{R}^Ty$ belongs to span$\{\mathbf{P}\}$, thus $\mathbf{P}\mathbf{R}^T - \mathbf{P}_\theta\mathbf{R}_\theta^T$ projects $\mathbf{y}$ onto the complement space of span$\{\mathbf{P}_\theta\}$ in span$\{\mathbf{P}\}$. The proof of $\mathbf{I}_m - \mathbf{P}\mathbf{R}^T - \mathbf{P}_r\mathbf{R}_r^T$ serving as the projector onto the complement subspace of span$\{\mathbf{P}_r\}$ in span$\{\mathbf{R}\}^\perp$ is similar, therefore omitted.

□

Based on the above conclusion, $\mathbb{S}_{\theta,1} =$ span$\{\mathbf{P}_\theta\}$, $\mathbb{S}_{\theta,2} =$ span$\{\mathbf{P}_{rr}\}$, $\mathbb{S}_{\theta^\perp,1} =$ span$\{\mathbf{P}_r\}$ and $\mathbb{S}_{\theta^\perp,2} =$ span$\{\mathbf{P}_o\}$ are achieved. On the whole, we can get $\mathbb{S}_\theta = \mathbb{S}_{\theta,1} \oplus \mathbb{S}_{\theta,2}$ and $\mathbb{S}_{\theta^\perp} = \mathbb{S}_{\theta^\perp,1} \oplus \mathbb{S}_{\theta^\perp,2}$. We introduce the following theorem for the C-PLS model:

Theorem 4.3. *For C-PLS model, it holds that*

$$\begin{cases}\mathbf{\Pi}_{\theta^\perp} = \mathbf{P}_{\theta^\perp}\mathbf{P}_{\theta^\perp}^T \\ \mathbf{\Pi}_\theta = \mathbf{I}_m - \mathbf{P}_{\theta^\perp}\mathbf{P}_{\theta^\perp}^T\end{cases} \tag{4.30}$$

Table 4.1: Summary of projectors

Projector	Direct	LS	PCR	PLS	T-PLS
$\mathbf{\Pi}_\theta$	$\mathbf{U}_{\theta\mathbf{y}}\mathbf{U}_{\theta\mathbf{y}}^T$	$\mathbf{Q}_1\mathbf{Q}_1{}^T$	$\mathbf{P}_{PCR}\mathbf{R}_{PCR}^T$	$\mathbf{PR}^T$	$\mathbf{I}_m - \mathbf{PR}^T - \mathbf{P}_r\mathbf{R}_r^T + \mathbf{P}_\theta\mathbf{R}_\theta^T$
$\mathbf{\Pi}_{\theta^\perp}$	$\mathbf{U}_{\theta\mathbf{y}^\perp}\mathbf{U}_{\theta\mathbf{y}^\perp}^T$	$\mathbf{Q}_2\mathbf{Q}_2{}^T$	$\mathbf{I}_m - \mathbf{P}_{PCR}\mathbf{R}_{PCR}^T$	$\mathbf{I}_m - \mathbf{PR}^T$	$\mathbf{PR}^T - \mathbf{P}_\theta\mathbf{R}_\theta^T + \mathbf{P}_r\mathbf{R}_r^T$
			C-PLS		
			$\mathbf{I}_m - \mathbf{P}_{\theta^\perp}\mathbf{P}_{\theta^\perp}^T$		
			$\mathbf{P}_{\theta^\perp}\mathbf{P}_{\theta^\perp}^T$		

Proof. The KPI-uncorrelated subspace is:

$$\mathbf{y}_{\theta^\perp} = \mathbf{P}_{\theta^\perp}\mathbf{R}_{\theta^\perp}^T\mathbf{y} = \mathbf{P}_{\theta^\perp}\mathbf{P}_{\theta^\perp}^T\left(\mathbf{I}_m - \mathbf{R}_\theta^\dagger\mathbf{R}_\theta^T\right)\mathbf{y} \tag{4.31}$$

Note that $\mathbf{P}_{\theta^\perp}^T\mathbf{R}_\theta = 0$ holds, since $\mathbf{P}_{\theta^\perp}$ belongs to the space spanned by $\mathbf{R}_\theta^\perp$. Thus $\mathbf{y}_{\theta^\perp}$ could be simply calculated as $\mathbf{y}_{\theta^\perp} = \mathbf{P}_{\theta^\perp}\mathbf{P}_{\theta^\perp}^T\mathbf{y}$, namely, $\mathbf{\Pi}_{\theta^\perp} = \mathbf{P}_{\theta^\perp}\mathbf{P}_{\theta^\perp}^T$. On the other hand, the KPI-correlated space is calculated by $\mathbf{y} - \mathbf{P}_{\theta^\perp}\mathbf{P}_{\theta^\perp}^T\mathbf{y}$, then the KPI-correlated part produced by the C-PLS model is $\mathbf{\Pi}_\theta = \mathbf{I}_m - \mathbf{P}_{\theta^\perp}\mathbf{P}_{\theta^\perp}^T$. □

Accordingly, $\hat{\mathbf{y}}_{C-PLS} = \left(\mathbf{I}_m - \mathbf{P}_{\theta^\perp}\mathbf{P}_{\theta^\perp}^T\right)\mathbf{y}$ consists of two parts, namely, $\mathbf{y}_\theta$ and $\tilde{\mathbf{y}}_\theta$. If $S_\theta = \text{span}\{\mathbf{P}_{\theta^\perp}^\perp\}$ is chosen as the KPI-correlated subspace, the space could also be represented using two separate subspaces, *i.e.* $\mathbb{S}_\theta = \mathbb{S}_{\theta,1} \oplus \mathbb{S}_{\theta,2}$ with $\mathbb{S}_{\theta,1} = \text{span}\{\mathbf{R}_\theta\}$ and $\mathbb{S}_{\theta,2} = \text{span}\{\mathbf{P}_{\tilde{\theta}}\}$. C-PLS models $\mathbf{y}$ as $\mathbf{y} = \mathbf{y}_{\hat{\theta}} + \mathbf{y}_{\theta^\perp} + \mathbf{y}_{\tilde{\theta}} = (\mathbf{R}_\theta^T)^\dagger\mathbf{R}_\theta^T\mathbf{y} + \mathbf{P}_{\theta^\perp}^T\mathbf{R}_{\theta^\perp}^T\mathbf{y} + \mathbf{y}_{\tilde{\theta}}$ and employs $\hat{\mathbf{y}}_{C-PLS} = \mathbf{y}_{\hat{\theta}} + \mathbf{y}_{\tilde{\theta}}$ for monitoring KPI-related faults, while $\tilde{\mathbf{y}}_{C-PLS} = \mathbf{y}_{\theta^\perp}$ serves for KPI-unrelated faults.

The two projectors $\mathbf{\Pi}_\theta$ and $\mathbf{\Pi}_{\theta^\perp}$ induced by all methods are summarized in Table 4.1. The subspaces information projected by the projectors, *i.e.*, $\mathbb{S}_\theta$ and $\mathbb{S}_{\theta^\perp}$ is shown in Table 4.2. From Tables 4.1 and 4.2, it can be observed that the projectors fulfil the condition $\mathbf{\Pi}_\theta + \mathbf{\Pi}_{\theta^\perp} = \mathbf{I}_m$ and the two subspace jointly span the full $\mathbf{y}$-space. The projectors' expressions show that the direct solution, LS and C-PLS approaches use orthogonal projections, while the others use oblique ones. In Table 4.2, the dimensionality of the $\mathbb{S}_\theta$ for the direct solution, LS and PCR-based methods equals l, namely the number of KPI variables. In the decomposition by PLS model, the KPI-correlated subspace has the dimension of κ, and $\kappa \geq l$ always holds, which shows that this part has redundant information responsible for KPI. For the T-PLS and C-PLS, $\dim\{S_\theta\} \geq l$.

Figure 4.1 shows the projections rendered by the direct method, we can see that the orthogonal projections performed on $\mathbf{y}$ are directly af-

Table 4.2: Information about KPI-correlated subspaces

	Direct	LS	PCR	PLS	T-PLS	C-PLS
$\mathbb{S}_\theta$	$\mathbf{U}_{\theta y}$	$\mathbf{Q}_1$	$\mathbf{P}_{PCR}$	$\mathbf{P}$	$\mathbf{P}_\theta \oplus \mathbf{P}_{rr}$	$\mathbf{P}_{\theta\perp}^{\perp}$
Dimension	l	l	l	κ	$\kappa_{rr} + l$	$m - \kappa_{\theta\perp}$

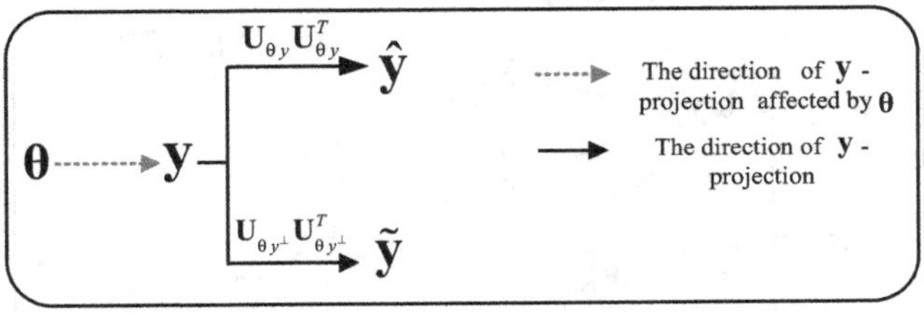

Figure 4.1: Demonstration of the projections of the direct method

fected by $\boldsymbol{\theta}$ (the cross-covariance between $\mathbf{y}$ and $\boldsymbol{\theta}$). This figure will serve as a reference for understanding the other methods. Figure 4.2 shows the projection schematics by LS and PCR based methods. It can be seen how the predicted $\hat{\boldsymbol{\theta}}$ affect the projections, in addition, the different ways they obtain the subspaces can be distinguished by the orthogonal and oblique projections. The projection connections between T-PLS and PLS are shown in Figure 4.3, where the dashed arrows indicate that the decomposition/projection about $\mathbf{y}$ is caused by the corresponding part in $\boldsymbol{\theta}$. From the figure, it could be well understood how T-PLS model could be obtained from PLS. Similar to Figure 4.3, Figure 4.4 shows the projection connections between CPLS and PLS. The orthogonal projections have been distinguished from the oblique ones in T-PLS. Also, it can also be concluded that C-PLS is not resultant from the deep decomposition of PLS, while a re-decomposition on $\mathbf{y}$ based on $\hat{\boldsymbol{\theta}}_{PLS}$.

Computations

Table 4.3 shows the computational complexity and the needed design effort (key parameters) for the methods, where it is supposed that the computation of PCA model involved in the methods under consideration is standardized with an SVD on the corresponding covariance matrix. The direct SVD based method requires only a single SVD on the $m \times l$

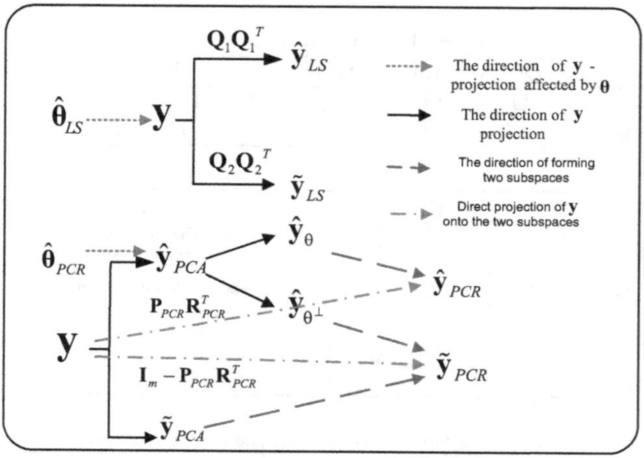

Figure 4.2: Demonstration of the projections of LS and PCR

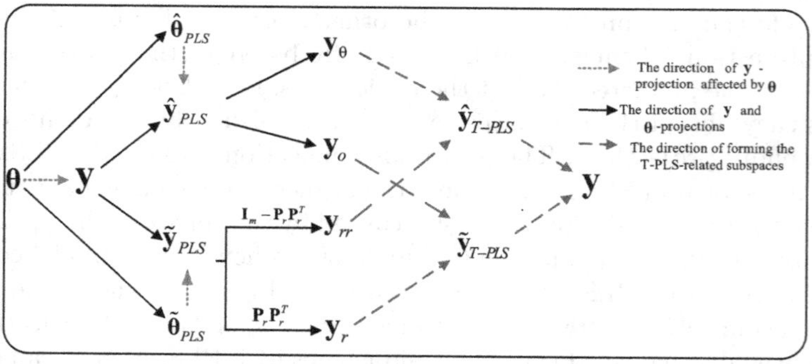

Figure 4.3: Demonstration of the projection relationship between PLS and T-PLS

cross-covariance matrix, besides, there is no parameter needed to design. LS-based methods include an SVD on the $m \times m$ covariance matrix and a QR decomposition on an $m \times l$ matrix, which are more costly than the direct one. PCR-based method involves the same computation with LS, but it includes an extra parameter $\bar{m}$ to specify. The pure PLS method needs κ eigenvalue decompositions on $l \times l$ matrices. Since $l << m$, the computational burden of PLS will significantly depend on the

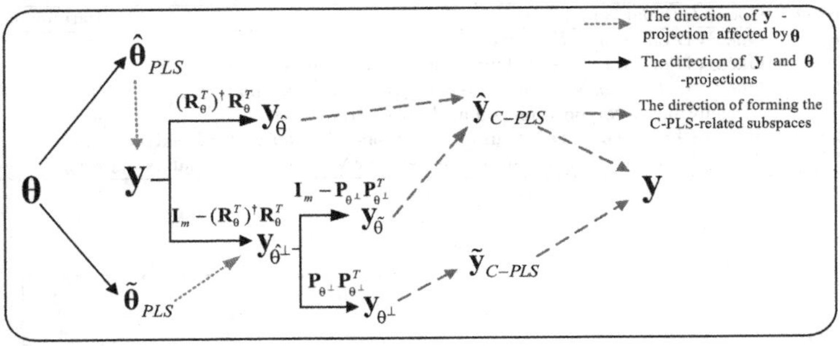

Figure 4.4: Demonstration of the projection relationship between PLS and C-PLS

parameter κ, which is a user-predefined parameter adopted in all PLS-related approaches, and can considerably effect the PM-FD performance. Although some certain criteria (*e.g.*, cross-validation) has been broadly used, the design of κ for the PM-FD purpose is still an open topic [50, 51]. The heaviest calculation burden can be found in T-PLS and C-PLS. T-PLS involves two more SVDs on an $m \times m$ matrix and one more SVD on an $l \times l$ matrix than the cost of PLS. C-PLS is a bit better, which has one SVD on an $m \times m$ matrix less than T-PLS. In detail, the flops attained for all methods are demonstrated in Figure 4.5, where the results are based on the experiments considering (1) without loss of generality, $l = 2$ is selected; (2) the unified SVD and QR decomposition are taken for use in all methods; (3) $\kappa = m/3$ is assumed. From the results, we can observe that the PLS method shows the minimal flops while the T-PLS method gives the largest one. However, if the determination cost of κ is added up, for example considering the 10-fold cross-validation [51], the PLS based cost will be higher than the direct one as shown in the Figure 4.5. With the increase of m, the cost of the linear regression-based methods will converge to the C-PLS. But C-PLS still concludes the design of κ, thus has more extra costs. Herein, it needs further stressed that the low efficiency of T-PLS actually comes from the subsequent PCA models while not PLS as commented in [40, 46].

Table 4.3: Summary of the computational complexity and parameter

Approach	Computational complexity	Parameter
Direct	One SVD on an $m \times l$ matrix	—
LS	One SVD on an $m \times m$ matrix +one QR on an $m \times l$ matrix	—
PCR	One SVD on an $m \times m$ matrix +one QR on an $m \times l$ matrix	$\bar{m}$
PLS	κ eigenvalue decompositions on an $l \times l$ matrix	κ
T-PLS	cost of PLS +two SVDs on an $m \times m$ + one SVD on an $l \times l$ matrix	κ, κ_r
C-PLS	cost of PLS+one SVD on an $m \times m$ + one SVD on an $l \times l$ matrix	$\kappa, \kappa_{\theta\perp}$

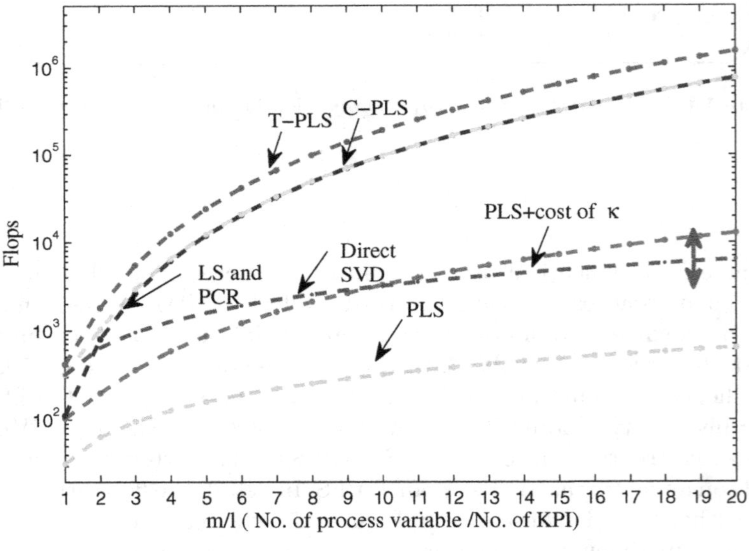

Figure 4.5: Flops costed by the examined methods

4.3.3 Remarks for PM-FD

To deal with PM-FD events, the direct solution successfully isolates the $\boldsymbol{\theta}$-correlated space from the $\boldsymbol{\theta}$-uncorrelated space with an SVD, based on which two separate PM-FD indices are developed. With the aid of linear regression tools, the LS and PCR- based methods divide $\mathbf{y}$ into the section directly responsible for predicting $\boldsymbol{\theta}$, and the section of complete indifference for $\boldsymbol{\theta}$. Since the two methods address the KPI-based PM-FD issue from distinct viewpoints, it is not reasonable to propose which is superior. Comparatively, the method based on PLS is totally different, where the KPI-correlated part ($\hat{\mathbf{y}}_{PLS}$) is obtained uncorrelated with the KPI-uncorrelated part, and also responsible for explaining the PLS-prediction of KPI ($\hat{\boldsymbol{\theta}}_{PLS}$). The statistical index is based on $\hat{\mathbf{y}}_{PLS}$ instead of $\hat{\boldsymbol{\theta}}_{PLS}$, which causes significant false alarms. T-PLS and C-PLS both place strong emphasis on $\hat{\boldsymbol{\theta}}_{PLS}$, and they reduce the false alarm rate. Furthermore, they also consider the fact that the KPI-uncorrelated part in PLS ($\tilde{\mathbf{y}}_{PLS}$) nevertheless may affect $\boldsymbol{\theta}$ due to the weak correlation with $\boldsymbol{\theta}$. Although it is unclear how this part affects KPI, there appears no doubt that concluding this part as the potentials would be favorable for the detection result. The improvements could also be represented by referring the $\mathbb{S}_\theta$ in Table 4.2.

The threshold setting for the Hotelling's T^2 statistic should be emphasized [52]. Comparing Eqs. (4.7) and (4.12), there are two different schemes for bounding the T^2 index [52], which have been discussed in Chapter 3. It it worth noting that from the implementation aspect, the difference between them is trivial when a sufficient amount of training data have been collected, while with limited data, the $\mathcal{F}$-distribution based one is more common. In the following performance evaluating part, Eq. (4.7) will be adopted so that the theoretical results in the last chapters can be used.

4.4 Performance evaluation

In this section, the KPI-related PM-FD approaches under consideration will be evaluated with respect to the EDD index proposed in Chapter 2. Firstly, the methods are reformulated in terms of how to apply the T^2 and Q statics for FD purpose. Secondly, based on the results shown in Chapters 2 and 3, the approach to calculate FDR and EDD will be given. A numerical example will be simulated to evaluate the methods by implementing the EDD index. It is firstly noted that when a fault occurs, it may not necessarily affect KPI. It is assumed that $\hat{\mathbf{y}}$ and $\tilde{\mathbf{y}}$ have been obtained by a certain approach, and the detection indices for the two parts have been designed, respectively. Then, the faulty $\mathbf{y}$ could be monitored by them and accordingly gives the following decisions: (1) if only $\hat{\mathbf{y}}$ part reports alarms, then $\boldsymbol{\theta}$ is affected; (2) if both parts have alarms, then $\boldsymbol{\theta}$ is also affected; (3) else if only $\tilde{\mathbf{y}}$ gives alarms, then $\boldsymbol{\theta}$ is unaffected. It can be drawn that, provided that there is no *a priori* faulty knowledge, the individual monitoring of $\hat{\mathbf{y}}$ parts can play an even important role to indicates KPI's status. Thus, this part will merely examine the performance of the detection indices for this part achieved by all methods.

4.4.1 A unified form of KPI-related fault detection

This section considers the KPI-based methods that only monitor the subspace $\mathbb{S}_\theta$, *i.e.* the $\boldsymbol{\theta}$-related subspace. The offline KPI dataset is again structured as $\boldsymbol{\Theta} = [\boldsymbol{\theta}_1, ..., \boldsymbol{\theta}_N] \in \mathbb{R}^{l \times N}$. The goal is to monitor $\boldsymbol{\theta}$ online using only the process measurement $\mathbf{y}$.

Methods only using the T^2-statistic

A general form for this type of method is

$$J_\theta = J_{\theta,T^2} = \mathbf{y}^T \mathbb{P}_\theta \boldsymbol{\Lambda}_\theta^{-1} \mathbb{P}_\theta^T \mathbf{y} \tag{4.32}$$

where $\mathbb{P}_\theta$ also denotes the basis vectors for the subspace, *i.e.* $\mathbb{S}_\theta = \text{span}\{\mathbb{P}_\theta\}$, $\boldsymbol{\Lambda}_\theta = \frac{\mathbb{P}_\theta^T \mathbf{Y}\mathbf{Y}^T \mathbb{P}_\theta}{N-1}$. As discussed in Section 4.3, four methods, *i.e.* direct, LS, PCR and PLS have such a form. $\mathbb{P}_\theta$ for them can be obtained by referring to Tables 4.1 and 4.2. The threshold is calculated using $\chi^2_{\kappa,\alpha}$ or $\chi^2_{l,\alpha}$.

Methods using both the T^2- and Q-statistics

Like in the PCA-based method for the complete measurement space, two statistics are developed in the form of

$$J_{\theta,T^2} = \mathbf{y}^T \bar{\mathbb{P}}_\theta \bar{\Lambda}_\theta^{-1} \bar{\mathbb{P}}_\theta^T \mathbf{y},\ J_{\theta,Q} = \mathbf{y}^T \tilde{\mathbb{P}}_\theta \tilde{\mathbb{P}}_\theta^T \mathbf{y} \tag{4.33}$$

where $\bar{\boldsymbol{\Lambda}}_\theta = \frac{\bar{\mathbb{P}}_\theta^T \mathbf{Y}\mathbf{Y}^T \bar{\mathbb{P}}_\theta}{N-1}$. In these methods, $\mathbb{S}_\theta = \text{span}\left\{\bar{\mathbb{P}}_\theta\right\} \oplus \text{span}\left\{\tilde{\mathbb{P}}_\theta\right\}$, $\dim\{\bar{\mathbb{P}}_\theta\} = \kappa$ [128]. The detection should consider both of them in a way of $J_\theta = J_{\theta,T^2} \vee J_{\theta,Q}$. C-PLS [24] and T-PLS [37] are included in this group, where $\bar{\mathbb{P}}_\theta$ and $\tilde{\mathbb{P}}_\theta$ are determined by consulting Tables 4.1 and 4.2. Thresholds are determined similarly to that discussed above.

4.4.2 Calculation of FDR for $J_{T^2,\mathbb{P}}$ and $J_{Q,\mathbb{P}}$

Let $J_{T^2,\mathbb{P}} = \mathbf{y}^T \mathbb{P} \Lambda_{\mathbb{P}}^{-1} \mathbb{P}^T \mathbf{y}$ and $J_{Q,\mathbb{P}} = \mathbf{y}^T \mathbb{P}\mathbb{P}^T \mathbf{y}$ be the generic expression for Eqs. (4.32) and (4.33). Let $\bar{\mathbf{y}} = \mathbb{P}^T \mathbf{y}$, $\breve{m} = \text{rank}(\mathbb{P}) \leq m$. When an additive fault occurs, *i.e.* $\mathbf{y}_f = \mathbf{y} + \Xi f$, $\bar{\mathbf{y}}_f = \mathbb{P}^T \mathbf{y} + \mathbb{P}^T \Xi f = \mathbb{P}^T \mathbf{y} + \Xi_{\mathbb{P}} f$. In this case, FDR is given by

$$\text{FDR} = \text{prob}\left(\chi^2_{\breve{m}}\left(\bar{\zeta} f^2\right) > \chi^2_{\breve{m},\alpha}\right) \tag{4.34}$$

where $\bar{\zeta} = \Xi_{\mathbb{P}}^T \Lambda_{\mathbb{P}}^{-1} \Xi_{\mathbb{P}}$. Similarly, FDR for $J_{Q,\mathbb{P}}$ is obtained by

$$\text{FDR} = \text{prob}\left\{\sum_{i=1}^{\breve{m}} \bar{\lambda}_i \chi^2_1\left(\bar{\delta}_i\right) > \sum_{i=1}^{\breve{m}} \bar{\lambda}_i \chi^2_{1,\alpha}\right\} \tag{4.35}$$

where $\bar{\lambda}_i$ denotes the ith eigenvalue of $\mathbf{\Lambda}_{\mathbb{P}}$, $\bar{\delta}_i = \bar{\Xi}_{\mathbb{P},i} f$, $\bar{\Xi}_{\mathbb{P}} = \mathbf{\Lambda}_{\mathbb{P}}^{-1} \mathbf{P}_{\mathbb{P}}^T \Xi_{\mathbb{P}}$, $\mathbf{P}_{\mathbb{P}}^T$ is the eigenvector of $\mathbf{\Lambda}_{\mathbb{P}}$. Eq. (4.35) can be approximated using the approach in Eq. (3.11).

When a multiplicative fault occurs, such as an independent multiplicative fault with $\mathbf{y}_f = \mathbf{M}\mathbf{y}$ and $\mathbf{M} = \text{diag}\left(M_1, \cdots, M_m\right) \forall i, M_i \geq 1$, it leads to $\bar{\mathbf{y}}_f = \mathbb{P}^T \mathbf{M} \mathbf{y}$. We can see that $\text{Var}(\bar{\mathbf{y}}_f)$ will also become larger under this kind of fault. To avoid estimating the magnitude of the change in $\bar{\mathbf{y}}_f$, the FDR can be calculated as

$$\begin{aligned} \text{FDR} &= \text{prob}\left(g_{\mathbb{P},f} \chi^2_{h_{\mathbb{P},f}} > J_{th,\mathbb{P}} = g_{\mathbb{P}} \chi^2_{h_{\mathbb{P}},\alpha}\right) \\ &= \text{prob}\left(\chi^2_{h_{\mathbb{P},f}} > \frac{g_{\mathbb{P}}}{g_{\mathbb{P},f}} \chi^2_{h_{\mathbb{P}},\alpha}\right) \end{aligned} \tag{4.36}$$

In the case, that $J_{T^2,\mathbb{P}}$ is adopted, then $g_{\mathbb{P}} = 1$, $h_{\mathbb{P}} = \breve{m}$, $g_{\mathbb{P},f} = \frac{\text{tr}\left(\mathbb{P}\Lambda_{\mathbb{P}}^{-1}\mathbb{P}^T \mathbf{M}\Sigma_y \mathbf{M}\mathbb{P}\Lambda_{\mathbb{P}}^{-1}\mathbb{P}^T \mathbf{M}\Sigma_y \mathbf{M}\right)}{\text{tr}\left(\mathbb{P}\Lambda_{\mathbb{P}}^{-1}\mathbb{P}^T \mathbf{M}\Sigma_y \mathbf{M}\right)}$ and $h_{\mathbb{P},f} = \frac{\text{tr}^2\left(\mathbb{P}\Lambda_{\mathbb{P}}^{-1}\mathbb{P}^T \mathbf{M}\Sigma_y \mathbf{M}\right)}{tr\left(\mathbb{P}\Lambda_{\mathbb{P}}^{-1}\mathbb{P}^T \mathbf{M}\Sigma_y \mathbf{M}\mathbb{P}\Lambda_{\mathbb{P}}^{-1}\mathbb{P}^T \mathbf{M}\Sigma_y \mathbf{M}\right)}$. In the case that $J_{Q,\mathbb{P}}$ is used, then $g_{\mathbb{P}} = \frac{\text{tr}\left(\mathbb{P}\mathbb{P}^T \Sigma_y \mathbb{P}\mathbb{P}^T \Sigma_y\right)}{\text{tr}(\mathbb{P}\mathbb{P}^T \Sigma_y)}$, $h_{\mathbb{P}} = \frac{\text{tr}^2\left(\mathbb{P}\mathbb{P}^T \Sigma_y\right)}{\text{tr}(\mathbb{P}\mathbb{P}^T \Sigma_y \mathbb{P}\mathbb{P}^T \Sigma_y)}$, $g_{\mathbb{P},f} = \frac{\text{tr}\left(\mathbf{M}\mathbb{P}\mathbb{P}^T \mathbf{M}\Sigma_y \mathbf{M}\mathbb{P}\mathbb{P}^T \mathbf{M}\Sigma_y\right)}{\text{tr}(\mathbf{M}\mathbb{P}\mathbb{P}^T \mathbf{M}\Sigma_y)}$ and $h_{\mathbb{P},f} = \frac{\text{tr}^2\left(\mathbf{M}\mathbb{P}\mathbb{P}^T \mathbf{M}\Sigma_y\right)}{\text{tr}(\mathbf{M}\mathbb{P}\mathbb{P}^T \mathbf{M}\Sigma_y \mathbf{M}\mathbb{P}\mathbb{P}^T \mathbf{M}\Sigma_y)}$. It is solved by integrating the PDF of $\chi^2_{h_{\mathbb{P},f}}$ over the region $\left[\ \frac{g_{\mathbb{P}}}{g_{\mathbb{P},f}} \chi^2_{h_{\mathbb{P}},\alpha} \quad \infty\ \right]$.

4.4.3 Simulation results

In this section, a numerical case is provided for demonstrating the theory. The simulated example without faults is

$$\begin{cases} \mathbf{y}_{obs}(k) = \mathcal{W}\mathbf{x}(k) + \boldsymbol{\nu}(k) \\ \boldsymbol{\theta}_{obs}(k) = \boldsymbol{\Psi}_{obs}\mathbf{y}_{obs}(k) + \mathbf{b}_{obs} + \boldsymbol{\zeta}(k) \end{cases} \tag{4.37}$$

where $\mathbf{y}_{obs} \in \mathbb{R}^m$, $\boldsymbol{\theta}_{obs} \in \mathbb{R}^l$ with $m = 6$, $l = 2$. $\mathbf{x} \in \mathbb{R}^3$ is simulated with three random sequences, *i.e.* $\mathbf{x}_1(k) \sim \mathcal{N}(0, 1)$, $\mathbf{x}_2(k) \sim \mathcal{N}(0, 0.5^2)$ and $\mathbf{x}_3(k) \sim \mathcal{N}(0, 1.5^2)$. The model parameters are generated in the forms of $\mathcal{W} = \text{randn}(6, 3)$, $\boldsymbol{\Psi}_{obs} = \text{randn}(2, 6)$, $\mathbf{b}_{obs} = [12, 10]^T$. To simulate the system uncertainties and measurement noise, two stochastic elements are added with $\boldsymbol{\nu}_i(k) \sim \mathcal{N}(0, 0.01^2)$ for $i = 1, ..., m$ and $\boldsymbol{\zeta}_j(k) \sim \mathcal{N}(0, 0.05^2)$ for $j = 1, ..., l$.

The faulty scenarios are simulated in the form of $\mathbf{y}_{obs,f} = \mathbf{y}_{obs} + \Xi f$, where Ξ denotes the direction/property of the fault and f is the fault magnitude. To show the holistic performance, both the KPI-related and KPI-unrelated faults shall be considered. As expected, the methods that offer a shorter EDD for the faults with KPI affected and a lengthy EDD or even cannot signal for the faults having no influence to KPI are seen as the high-score ones. Two types of faults can be realized by flexibly setting Ξ. We first perform a QR decomposition on $\boldsymbol{\Psi}_{obs}^T$:

$$\boldsymbol{\Psi}_{obs}^T = \mathbf{Q}_\Psi \mathbf{R}_\Psi = \left[\mathbf{q}_\Psi^{(1)}, \mathbf{q}_\Psi^{(2)}, ..., \mathbf{q}_\Psi^{(6)}\right] \begin{bmatrix} \mathbf{r}_\Psi^{(1)} \\ \mathbf{r}_\Psi^{(2)} \\ 0 \end{bmatrix}$$

Then the KPI-related fault direction basis is configured as $\Xi_\theta = \{\mathbf{q}_\Psi^{(1)}, \mathbf{q}_\Psi^{(2)}\}$. The KPI-unrelated fault direction could be chosen from the set $\Xi_{\theta\perp} = \{\mathbf{q}_\Psi^{(3)}, \mathbf{q}_\Psi^{(4)}, \mathbf{q}_\Psi^{(5)}, \mathbf{q}_\Psi^{(6)}\}$. To give a fair comparison, in this part, three faulty episodes respectively for both types of faults. In addition, the corresponding fault magnitude f is also changed with three different values, *i.e.* 0.5, 1.5 and 3.

For the KPI-related fault

Table 4.4 shows the detailed performance with EDD index for KPI-related faults. When Ξ_θ is configured as $\mathbf{q}_\Psi^{(1)}$, with the increase of f,

Table 4.4: EDD for different KPI-related faults for the numerical example

Ξ_θ	f	Direct	LS	PCR	PLS	T-PLS	C-PLS
	0.5	10.8370	0.0851	0.8430	0.0010	0.0020	0.0010
$\mathbf{q}_\Psi^{(1)}$	1.5	3.7766	0	0.5723	0	0	0
	3	0.4604	0	0.0639	0	0	0
	0.5	4.0042	3.7374	3.1732	2.1269	1.3612	3.0124
$\mathbf{q}_\Psi^{(2)}$	1.5	0.5580	0.4863	0.3418	0.1239	0.0564	0.1063
	3	0.0561	0.0423	0.0195	0.0014	0.0015	0
	0.5	6.0688	6.8403	1.7361	0.1925	0.1474	0.2657
$\frac{\sqrt{2}}{2}\mathbf{q}_\Psi^{(1)} + \frac{\sqrt{2}}{2}\mathbf{q}_\Psi^{(2)}$	1.5	1.1543	1.4711	0.0744	0	0.0337	0
	3	0.2070	0.3128	0.0004	0	0	0

the EDD values decreas for all methods. While f is small enough (*e.g.*, 0.5), the direct solution and PCR provide too large an EDD to detect it. For a larger f, PLS-based methods show smaller EDD compared with those based on SVD and LS, of which T-PLS and C-PLS are especially preferred. In the cases of other fault directions, the results are identical with those given by the first one. Thus, the ability of all considered methods in this case can be ranked as: PLS, linear regression and the direct one.

For the KPI-unrelated fault

The KPI-unrelated faulty scenarios are simulated with the results shown in Table 4.5. Since the faults are unrelated with KPI, the better performance delivered by the approach should have a large EDD. In this case, it is observed that the linear regression based methods give the best performance, because in most faulty cases, they can not detected the faults, in other words they have provided the least the wrong detections. Moreover, PLS based methods show the smallest EDD for any scenarios, which suggests that this kind of method has the highest probability of a false alarm. Within the PLS family, T-PLS and C-PLS present less EDD than PLS, which means they suffer a higher risky for false alarms. Therefore, it is concluded that the PLS based methods are not appropriate when a KPI-unrelated fault occurs, but those based on linear regression are more reliable.

Table 4.5: EDD for KPI-unrelated faults in numerical example

Ξ_θ	f	Direct	LS	PCR	PLS	T-PLS	C-PLS
	0.5	—	11.0766	—	8.0659	7.0210	11.0555
$\mathbf{q}_\Psi^{(3)}$	1.5	14.4003	4.6851	—	2.0528	1.0203	1.0000
	3	11.2532	1.9624	14.5771	0.5139	0.0125	0.0943
	0.5	—	17.6811	14.7215	1.9261	0.9945	0.8547
$\mathbf{q}_\Psi^{(4)}$	1.5	14.2545	14.5531	7.7411	0.1046	0.1254	0.7648
	3	10.39341	11.1166	3.7022	0.0010	0.0015	0.2112
	0.5	17.0581	12.2077	17.9925	2.1858	3.1243	2.5510
$\mathbf{q}_\Psi^{(5)}$	1.5	—	14.7968	14.9843	0.0911	0.0716	0.0592
	3	12.1279	11.8453	11.5708	0.0004	0	0

Performance evaluation for multiplicative faults

To better examine their performance for detecting multiplicative and drift faults, three parameters in Eq. (4.37) are specified as [128]

$$\mathcal{W} = \begin{bmatrix} -0.2478 & 1.2591 & 1.5909 \\ 1.1816 & -0.7621 & -0.4943 \\ 0.5611 & 0.1295 & -0.1046 \\ -0.2014 & -3.1881 & 0.4801 \\ 1.9874 & 0.8019 & -0.2836 \\ 2.0540 & 0.1096 & -0.6684 \end{bmatrix}, \mathbf{\Psi}_{obs} = \begin{bmatrix} 1.8193 & 1.4053 \\ -0.5958 & -0.3885 \\ 0.0571 & 0.6382 \\ 0.5205 & -0.0510 \\ -1.4298 & -1.9791 \\ -1.3206 & 1.8934 \end{bmatrix}^T$$

According to Chapters 2 and 3, the multiplicative fault is modelled as $\mathbf{y}_f = \mathbf{My}$. In this experiment, six faults are simulated with each one uniformly located at an individual variable. M_i is set to be 2, 5, and 10 for $i = 1, .., m$. Note from $\mathbf{\Psi}_{obs}$ that $\mathbf{y}_1$, $\mathbf{y}_5$ and $\mathbf{y}_6$ are closely correlated with KPI, thus, the faults occurring in these variables are also used to evaluate the KPI-based methods. Calculations of FDR are referred to Eqs. (4.35), (4.36), (3.9) and (3.11). EDD is calculated using Eq. (2.15). To eliminate randomness, each experiment is repeated 100 times with different values for the random components and the mean value is obtained.

Table 4.6 shows the results. Firstly, it is noted that for all methods, as M_i increases, the EDD values decrease, which is expected given the results in Chapter 2. The results given by KPI-based methods for detecting KPI-related faults are presented in Table 4.6. It can be observed that the SVD and PCR-based methods performed poorly due to

Table 4.6: EDD for different multiplicative faults

Ξ	f	SVD	LS	PCR	PLS	T-PLS	C-PLS
	2	1.5595	0.5324	1.5972	1.0710	0.0045	1.1099
$\boldsymbol{\xi}_1$ [1]	5	0.5787	0.1980	0.5893	0.3986	0.0018	0.4136
	10	0.2792	0.0956	0.2834	0.1910	0.0009	0.1991
	2	7.0889	1.8491	9.1278	2.4877	0.0032	2.7448
$\boldsymbol{\xi}_5$	5	2.7380	0.3787	3.9206	0.8714	0.0013	0.9909
	10	1.2352	0.1567	1.7492	0.4138	0.0006	0.4720
	2	5.8738	0.6590	6.4983	2.6253	0.0028	3.3042
$\boldsymbol{\xi}_6$	5	2.1703	0.2224	2.4569	0.9115	0.0011	0.1530
	10	0.9932	0.1038	1.1174	0.4312	0.0006	0.5314

[1] $\boldsymbol{\xi}_i$ denotes the ith column of $\mathbf{I}_6$.

the large EDD values in all three fault scenarios. They are followed by LS and PLS and C-PLS-based methods, which give the best EDD. The EDD values show that T-PLS is the best method for these faults. The EDD values are close to zero which implies that this method can detect faults almost as quickly as the change in the variables.

Performance evaluation for drift faults

The drift fault is generated in the form of $\mathbf{y}_f(k) = \Xi f(k)$ with $f(k) = \rho k$, ρ denoting the slope of the fault, which is similar to Section 2.4. The fault direction, Ξ, is selected as one of the left singular vector of $\boldsymbol{\Psi}^T$, which resembles the approach used in the earlier work [71]. For each run of the simulation, the criterion for determining the stopping time instance k_s is to check if $\text{FDR}(k_s) = 1$. Based on $\text{FDR}(k), k = 1, ..., k_s$ calculated using Eqs. (4.36) and (4.37), the EDD is then obtained by Eq. (2.15). Table 4.7 shows the detailed EDD performance for drift faults. Note by definition that the first three faults are KPI-related, thus, will be simultaneously applied to assess the KPI-based methods. It is expected that as the slope of the fault increases, the EDD values will likewise decrease for all faults. There commonly exists delay (EDD > 0) for detecting this kind of fault due to the small magnitude in the incipient stage. Performance of the KPI-based methods was examined using EDD. It can be found that the SVD-, PCR-, and PLS-based methods perform similarly for all fault cases. They have similar EDDs, which are larger

Table 4.7: EDD for different drift faults

Ξ	ρ	SVD	LS	PCR	PLS	T-PLS	C-PLS
$\boldsymbol{\xi}_1$ [1]	0.1	8.1404	2.1949	8.1352	8.5962	3.4943	3.3037
	0.5	5.3479	1.0505	5.3437	5.5619	1.9120	1.7890
	1	4.0169	0.5915	4.0134	4.2925	1.1254	1.1460
$\boldsymbol{\xi}_2$	0.1	7.1454	3.0327	7.3423	6.3102	2.5347	2.2314
	0.5	4.5617	1.6009	4.7148	3.9025	1.2715	1.1650
	1	3.3640	1.0059	3.4903	2.8147	0.7548	0.6756
$\frac{\sqrt{2}}{2}\boldsymbol{\xi}_1 + \frac{\sqrt{2}}{2}\boldsymbol{\xi}_2$	0.1	7.9734	6.8704	7.6475	6.8020	2.3248	2.2837
	0.5	5.2138	4.3497	4.9546	4.2741	1.1520	1.1251
	1	3.9048	3.1898	3.6889	3.1168	0.6722	0.6521

[1] $\boldsymbol{\xi}_i$ denotes the ith left eigenvector of $\boldsymbol{\Psi}_{obs}^T$.

than for the LS-, T-PLS- and C-PLS-based methods. In addition, SVD is the poorest one, PLS is relatively better. T-PLS and C-PLS are also similar in performance in terms of EDD. The values obtained by them are better than LS, especially for the 2nd and 3rd faults. This suggests that, despite the popularity of KPI-based approaches, for drift faults LS-, T-PLS-, and C-PLS-based methods are preferable.

4.5 Conclusions

In this chapter, the KPI-based MSPM approaches for linear static processes are reviewed. The existing approaches have been, for the first time, sorted into three classes based on their mechanisms of how to obtain the KPI-correlated and -uncorrelated subspaces from the space spanned by the process variables. The three categories are the direct solution, LS-based and PLS-based methods. Based on the theoretical analysis, their interconnections and the costed computations have been uncovered, and furthermore, to evaluate their performance in detecting additive, multiplicative and drift faults, the EDD index has been extended to MSPM methods and fully applied in the simulation study. Finally, it should be further noted that (1) This chapter offered a unified platform to investigate and evaluate the performance of the examined methods. Probably there is no such thing as an always best method in this area since there is likely an interaction between methods and the properties of the applied industrial contexts; (2) Although PLS-based methods are usually preferred in the literature, the PLS model was not originally developed for KPI-based PM-FD. The pioneering use was created by scholars who pay more attention on formulating innovative solutions to meaningful engineering problems rather than spending time in the development of elegant mathematical solutions [16, 17, 19, 20]. Therefore, methods such as T-PLS and C-PLS were developed to address the specific issues in PLS model to achieve theoretical improvements. However, the new formulation causes increase in the design efforts, and would somehow bring extra drawbacks due to the inappropriate modifications. Also, rather than the post-modification of PLS, the pre-modification like LS-based methods seems cost-efficient and direct.

5 KPI-based PM-FD methods for steady-state dynamic processes: Two DPLS-based methods

Practically, it is common that $\boldsymbol{\theta}$ and $\mathbf{y}$ are dynamically associated with each other [114]. The next two chapters will introduce the development of methods that take into the dynamics between them. Firstly, according to the descriptions for dynamic processes in Chapter 2, when dynamic processes reach the steady state, it can be assumed that the mean and covariance of state and output variables keep stable, namely $\lim_{k\to\infty} \mathbf{x}(k) = \mathrm{E}(\mathbf{x}) = \mu_x$, $\lim_{k\to\infty} \mathbf{y}(k) = \mathrm{E}(\mathbf{y}) = \mu_y$, $\lim_{k\to\infty} \mathrm{Cov}(\mathbf{x}(k)) = \Sigma_x$ and $\lim_{k\to\infty} \mathrm{Cov}(\mathbf{y}(k)) = \Sigma_y$ [81, 82]. Since KPIs are always kept stable during the whole operation, they likewise satisfy that $\lim_{k\to\infty} \boldsymbol{\theta}(k) = \mathrm{E}(\boldsymbol{\theta}) = \mu_\theta$. In such case, the dynamic relation between KPI and process variables can be modeled using DPLS models, which lead to the development of DPLS-based PM-FD methods. Note that a commonly used DPLS method builds the model between past $\mathbf{y}$, $\boldsymbol{\theta}$ as well as current $\mathbf{y}$ and current $\boldsymbol{\theta}$, which is referred to as the auto-regressive moving average (ARMA) model. However, KPI-based models cannot be developed in such way as there is no online KPI measurement available. Without considering the past $\boldsymbol{\theta}$ term, the finite impulse response (FIR) model has been extensively adopted [12, 94]. This chapter takes into account the most popular two of them and compare them from both theoretical prospective and the performance of EDD.

5.1 Background

Consider a process with routine operating data

$$\underbrace{\begin{matrix}\mathbf{y}_{obs}(1), ..., \mathbf{y}_{obs}(k), ..., \mathbf{y}_{obs}(K) \\ \boldsymbol{\theta}_{obs}(1), ..., \boldsymbol{\theta}_{obs}(k), ..., \boldsymbol{\theta}_{obs}(K)\end{matrix}}_{K \text{ measurements}}$$

where $\mathbf{y}_{obs}(k) \in \mathbb{R}^m$ is the process measurement vector and $\boldsymbol{\theta}_{obs}(k) \in \mathbb{R}^l$ is the KPI vector. Assume that the process runs in steady state. Thus, the mean values of them can be removed, which, like the case in Chapter 4, gives the normalized data $\mathbf{y}$ and $\boldsymbol{\theta}$ for further use. Let k be the current time instance, and define

$$\mathbf{y}_s(k) = \begin{bmatrix} \mathbf{y}(k) \\ \vdots \\ \mathbf{y}(k-s+2) \\ \mathbf{y}(k-s+1) \end{bmatrix} \in \mathbb{R}^{sm}$$

$$\mathbf{Y}_{(k)} = [\mathbf{y}(k-N+1), \cdots, \mathbf{y}(k-1), \mathbf{y}(k)] \in \mathbb{R}^{m \times N}$$

and

$$\mathbf{Y}_g = [\mathbf{y}_s(k-N+1), \mathbf{y}_s(k-N+2), ..., \mathbf{y}_s(k)] = \begin{bmatrix} \mathbf{Y}_{(k)} \\ \mathbf{Y}_{(k-1)} \\ \vdots \\ \mathbf{Y}_{(k-s+1)} \end{bmatrix} \in \mathbb{R}^{sm \times N}$$

For the sake of simplification, rewrite $\mathbf{Y}_g$ as $\mathbf{Y}_g = \begin{bmatrix} \mathbf{Y}_{(0)} \\ \mathbf{Y}_{(1)} \\ \vdots \\ \mathbf{Y}_{(s-1)} \end{bmatrix}$. It represents the current and past process data. Similarly, the current KPI data can be structured as $\boldsymbol{\Theta} \in \mathbb{R}^{l \times N}$, which is the shorthand for $\boldsymbol{\Theta}_{(0)}$. Based on $\mathbf{Y}_g$ and $\boldsymbol{\Theta}$, the DPLS methods attempt to extract the information in $\mathbf{y}$ that can accurately explain $\boldsymbol{\theta}$, and allow a convenient implementation of the resulting model for online use.

Let $\mathbf{w}_{(j)} \in \mathbb{R}^m$ $j = 0, ..., s-1$ be the direction vector such that $\mathbf{t}_{(j)} = \mathbf{w}_{(j)}^T \mathbf{Y}_{(j)}$, $\boldsymbol{\beta} = [\boldsymbol{\beta}(0), \cdots, \boldsymbol{\beta}(s-1)]^T \in \mathbb{R}^s$ be the weighting vector that combines $\mathbf{t}_{(j)}$ as $\mathbf{t}_g = \sum_{j=0}^{s-1} \boldsymbol{\beta}(j)\mathbf{t}_{(j)}$. The objective function addressed by the DPLS method is

$$\begin{aligned} &\max_{\mathbf{w}_{(j)},\boldsymbol{\beta}} \left\|\mathbf{t}_g \boldsymbol{\Theta}^T\right\|_{\mathrm{E}}^2 \\ s.t. \quad &\left\|\mathbf{w}_{(j)}\right\|_{\mathrm{E}} = \|\boldsymbol{\beta}\|_{\mathrm{E}} = 1 \end{aligned} \tag{5.1}$$

Compared with the PLS model as shown in Section 4.2.3, Eq. (5.1) is an extension of the objective function of static PLS model.

5.2 A comparison of two DPLS models

This section introduces in more details two DPLS methods, including their original forms and the NIPALS alternatives. The comparisons and interconnections between them in mathematical formulations will be explicitly followed.

5.2.1 Two DPLS methods

There exist two methods for solving Eq. (5.1). The first method, which will be called the DDPLS method, solves Eq. (5.1) directly using Lagrange multipliers. The second method, which will be called the IDPLS, first assumes that $\mathbf{w}_{(j)} = \mathbf{w}\ \forall j$, and, then applies Lagrange multipliers to solve the simplified objective function. Noted that $\mathbf{w}$ has a similar definition with that defined in PLS model. The difference is that DPLS considers time-lagged process data.

The DDPLS method

In this method, $\mathbf{w}_{(j)}$ has no additional constraint, which implies that the KPI-relevant information in the process data blocks is extracted along their unique directions to make it closely correlated to the current KPI data. To solve this problem, applying the Lagrange multipliers to Eq.

Table 5.1: Original algorithm for the DDPLS method

1. Initialize $\mathbf{w}_{(j)}$ to be random unit vectors.
2. Take $\boldsymbol{\beta}$ as the eigenvector of S_w corresponding to the largest eigenvalue.
3. Let $\mathbf{w}_{(j)}$ be $\frac{\mathbf{S}_{\boldsymbol{\beta}(j)}}{\left\|\mathbf{S}_{\boldsymbol{\beta}(j)}\right\|_E}$
4. Return to Step 2 as long as $\boldsymbol{\beta}$ and $\mathbf{w}_{(j)}$ have not converged, otherwise, stop the algorithm.

(5.1) gives

$$\mathcal{L}(\lambda_{\mathbf{w}_{(j)}}, \lambda_{\boldsymbol{\beta}}) = \boldsymbol{\beta}^T \begin{bmatrix} \mathbf{w}_{(0)}^T \mathbf{Y}_{(0)} \\ \vdots \\ \mathbf{w}_{(s-2)}^T \mathbf{Y}_{(s-2)} \\ \mathbf{w}_{(s-1)}^T \mathbf{Y}_{(s-1)} \end{bmatrix} \boldsymbol{\Theta}^T \boldsymbol{\Theta} \left[\mathbf{Y}_{(0)}^T \mathbf{w}_{(0)}, \cdots \mathbf{Y}_{(s-2)}^T \mathbf{w}_{(s-2)}, \quad \mathbf{Y}_{(s-1)}^T \mathbf{w}_{(s-1)} \right] \boldsymbol{\beta}$$
$$-\lambda_{\mathbf{w}_{(0)}} \left(\mathbf{w}_{(0)}^T \mathbf{w}_{(0)} - 1\right) -, \cdots, -\lambda_{\mathbf{w}_{(s-1)}} \left(\mathbf{w}_{(s-1)}^T \mathbf{w}_{(s-1)} - 1\right) - \lambda_{\boldsymbol{\beta}} \left(\boldsymbol{\beta}^T \boldsymbol{\beta} - 1\right) \tag{5.2}$$

Taking the derivative of Eq. (5.2) with respect to $\mathbf{w}_{(j)}$ and $\boldsymbol{\beta}$, and setting them equal to zero gives

$$\underbrace{\begin{bmatrix} \mathbf{w}_{(0)}^T \mathbf{Y}_{(0)} \\ \vdots \\ \mathbf{w}_{(s-2)}^T \mathbf{Y}_{(s-2)} \\ \mathbf{w}_{(s-1)}^T \mathbf{Y}_{(s-1)} \end{bmatrix} \boldsymbol{\Theta}^T \boldsymbol{\Theta} \left[\mathbf{Y}_{(0)}^T \mathbf{w}_{(0)}, \quad \cdots, \quad \mathbf{Y}_{(s-2)}^T \mathbf{w}_{(s-2)}, \quad \mathbf{Y}_{(s-1)}^T \mathbf{w}_{(s-1)} \right]}_{\mathbf{S}_w} \boldsymbol{\beta} = \lambda_{\boldsymbol{\beta}} \boldsymbol{\beta}$$
$$\underbrace{\boldsymbol{\beta}(0)\, \mathbf{Y}_{(0)} \boldsymbol{\Theta}^T \boldsymbol{\Theta} \left[\mathbf{Y}_{(0)}^T \mathbf{w}_{(0)}, \quad \cdots, \quad \mathbf{Y}_{(s-2)}^T \mathbf{w}_{(s-2)}, \quad \mathbf{Y}_{(s-1)}^T \mathbf{w}_{(s-1)} \right] \boldsymbol{\beta}}_{\mathbf{S}_{\boldsymbol{\beta}(0)}} = \lambda_{\mathbf{w}_{(0)}} \mathbf{w}_{(0)}$$
$$\vdots$$
$$\underbrace{\boldsymbol{\beta}(s-1)\, \mathbf{Y}_{(s-1)} \boldsymbol{\Theta}^T \boldsymbol{\Theta} \left[\mathbf{Y}_{(0)}^T \mathbf{w}_{(0)}, \quad \cdots, \quad \mathbf{Y}_{(s-2)}^T \mathbf{w}_{(s-2)}, \quad \mathbf{Y}_{(s-1)}^T \mathbf{w}_{(s-1)} \right] \boldsymbol{\beta}}_{\mathbf{S}_{\boldsymbol{\beta}(s-1)}} = \lambda_{\mathbf{w}_{(s-1)}} \mathbf{w}_{(s-1)} \tag{5.3}$$

The optimal $\mathbf{w}_{(j)}$ and $\boldsymbol{\beta}$ can be achieved using the algorithm in Table 5.1. This algorithm converges since $0 < \mathcal{L} \le \lambda_{max}(\mathbf{Y}_g \boldsymbol{\Theta}^T \boldsymbol{\Theta} \mathbf{Y}_g^T)$. For the ith iteration, $\mathcal{L}^i \le \mathcal{L}^i(\boldsymbol{\beta}^{i+1}, \mathbf{w}_{(0)}^i, ..., \mathbf{w}_{(s-1)}^i) \le \mathcal{L}^i(\boldsymbol{\beta}^{i+1}, \mathbf{w}_{(0)}^{i+1}, ..., \mathbf{w}_{(s-1)}^i) \le ... \le \mathcal{L}^i(\boldsymbol{\beta}^{i+1}, \mathbf{w}_{(0)}^{i+1}, ..., \mathbf{w}_{(s-1)}^{i+1}) = \mathcal{L}^{i+1}$ holds.

After convergence has been achieved, the optimal $\mathcal{L}$ can be expressed as

$$\mathcal{L}_{max} = \lambda_{\boldsymbol{\beta}} = \sum_{j=0}^{s-1} \lambda_{\mathbf{w}_{(j)}} \tag{5.4}$$

where $\lambda_{\mathbf{w}(j)}$ equals to $\left\|\mathbf{S}_{\boldsymbol{\beta}(j)}\right\|_{\mathrm{E}}$ and $\lambda_{\boldsymbol{\beta}}$ is the largest eigenvalue of $\mathbf{S}_{\mathbf{w}}$.

The IDPLS method

In this method, $\mathbf{w}_{(j)} = \mathbf{w}\ \forall j$ is assumed, which implies that the KPI-relevant information will be captured from the process data blocks along a common direction. Using this assumption, the objective function (5.1) can be simplified as

$$\begin{array}{ll} \max\limits_{\mathbf{w},\boldsymbol{\beta}} & \left\|(\boldsymbol{\beta} \otimes \mathbf{w})^T \mathbf{Y}_g \boldsymbol{\Theta}^T\right\|_{\mathrm{E}}^2 \\ s.t. & \|\mathbf{w}\|_{\mathrm{E}} = \|\boldsymbol{\beta}\|_{\mathrm{E}} = 1 \end{array} \tag{5.5}$$

Applying the Lagrange multipliers-based approach gives

$$\mathcal{L}(\mathbf{w}, \boldsymbol{\beta}, \lambda_w, \lambda_{\boldsymbol{\beta}}) = \left\|(\boldsymbol{\beta} \otimes \mathbf{w})^T \mathbf{Y}_g \boldsymbol{\Theta}^T\right\|_{\mathrm{E}}^2 - \lambda_w(\mathbf{w}^T\mathbf{w} - 1) - \lambda_{\boldsymbol{\beta}}(\boldsymbol{\beta}^T\boldsymbol{\beta} - 1) \tag{5.6}$$

Taking the derivative of Eq. (5.6) with respect to $\boldsymbol{\beta}$ and $\mathbf{w}$ gives [12]

$$\begin{aligned} (\boldsymbol{\beta} \otimes \mathbf{I}_m)^T \mathbf{Y}_g \boldsymbol{\Theta}^T \boldsymbol{\Theta} \mathbf{Y}_g^T (\boldsymbol{\beta} \otimes \mathbf{I}_m) \mathbf{w} = \lambda_w \mathbf{w} \\ (\mathbf{I}_s \otimes \mathbf{w})^T \mathbf{Y}_g \boldsymbol{\Theta}^T \boldsymbol{\Theta} \mathbf{Y}_g^T (\mathbf{I}_s \otimes \mathbf{w}) \boldsymbol{\beta} = \lambda_{\boldsymbol{\beta}} \boldsymbol{\beta} \end{aligned} \tag{5.7}$$

From Eq. (5.7), we can see that $\mathbf{w}$ and $\boldsymbol{\beta}$ depend on each other. To obtain their optimal values, an iterative algorithm, given in Table 5.2, has been proposed [12].

Like for the DDPLS method, the convergence of the algorithm in Table 5.2 can be guaranteed since $0 < \mathcal{L} \leq \lambda_{max}(\mathbf{Y}_g \boldsymbol{\Theta}^T \boldsymbol{\Theta} \mathbf{Y}_g^T)$, $\lambda_{max}(\mathbf{Y}_g \boldsymbol{\Theta}^T \boldsymbol{\Theta} \mathbf{Y}_g^T)$ is the largest eigenvalue of $\mathbf{Y}_g \boldsymbol{\Theta}^T \boldsymbol{\Theta} \mathbf{Y}_g^T$, and for the ith iteration [93], $\mathcal{L}^i \leq \mathcal{L}^i(\boldsymbol{\beta}^i, \mathbf{w}^{i+1}) \leq \mathcal{L}^i(\boldsymbol{\beta}^{i+1}, \mathbf{w}^{i+1}) = \mathcal{L}^{i+1}$ is satisfied. After converging, the optimal $\mathcal{L}$ is

$$\begin{aligned} \mathcal{L}_{\max} &= \boldsymbol{\beta}^T (\mathbf{I}_s \otimes \mathbf{w})^T \mathbf{Y}_g \boldsymbol{\Theta}^T \boldsymbol{\Theta} \mathbf{Y}_g^T (\mathbf{I}_s \otimes \mathbf{w}) \boldsymbol{\beta} = \lambda_{\boldsymbol{\beta}} \\ &= \mathbf{w}^T (\boldsymbol{\beta} \otimes \mathbf{I}_m)^T \mathbf{Y}_g \boldsymbol{\Theta}^T \boldsymbol{\Theta} \mathbf{Y}_g^T (\boldsymbol{\beta} \otimes \mathbf{I}_m) \mathbf{w} = \lambda_w \end{aligned} \tag{5.8}$$

Table 5.2: Original algorithm for the IDPLS method

1. Randomly initialize $\boldsymbol{\beta}$ to be a unit vector.
2. Let $\mathbf{w}$ be the eigenvector of $(\boldsymbol{\beta} \otimes \mathbf{I}_m)^T \mathbf{Y}_g \boldsymbol{\Theta}^T \boldsymbol{\Theta} \mathbf{Y}_g^T (\boldsymbol{\beta} \otimes \mathbf{I}_m)$ corresponding to the largest eigenvalue.
3. Let $\boldsymbol{\beta}$ be the eigenvector of $(\mathbf{I}_s \otimes \mathbf{w})^T \mathbf{Y}_g \boldsymbol{\Theta}^T \boldsymbol{\Theta} \mathbf{Y}_g^T (\mathbf{I}_s \otimes \mathbf{w})$ corresponding to the largest eigenvalue
4. Return to Step 2 as long as both $\boldsymbol{\beta}$ and $\mathbf{w}$ have not converged, otherwise, stop the algorithm.

Table 5.3: NIPALS algorithm for the DDPLS method

1. Initialize $\mathbf{u} = \boldsymbol{\Theta}(1,:)$, and randomly select $\boldsymbol{\beta}$
2. $\mathbf{w}_{(j)} = \mathbf{Y}_{(j)}\mathbf{u}^T / \left\|\mathbf{Y}_{(j)}\mathbf{u}^T\right\|_{\mathrm{E}}$, $j = 0, .., s-1$
3. $\mathbf{t}_{(j)} = \mathbf{w}_{(j)}^T \mathbf{Y}_{(j)}$, $j = 0, .., s-1$
4. $\mathbf{t}_G = \left[\mathbf{t}_{(0)}^T, \cdots, \mathbf{t}_{(s-2)}^T, \mathbf{t}_{(s-1)}^T\right]^T$
5. $\boldsymbol{\beta} = \mathbf{t}_G \mathbf{u}^T / \left\|\mathbf{t}_G \mathbf{u}^T\right\|_{\mathrm{E}}$
6. $\mathbf{t}_g = \boldsymbol{\beta}^T \mathbf{t}_G$
7. $\mathbf{q} = \boldsymbol{\Theta}\mathbf{t}_g^T / \mathbf{t}_g \mathbf{t}_g^T$
8. $\mathbf{u} = \mathbf{q}^T \boldsymbol{\Theta} / \mathbf{q}^T \mathbf{q}$
9. Return to Step 2 until the convergence is reached.

5.2.2 The NIPALS alternative

The NIPALS method was first proposed by Wold [92] to iteratively calculate the principal components in PCA and PLS. It was proved to be equivalent to the eigenvalue decomposition-based method. This method can avoid performing matrix decomposition by replacing them with matrix multiplication. This section will introduce the NIPALS into the two DPLS methods.

If $\mathbf{Y}_{(j)}$ $\forall j$ are treated as s multiblock data, the algorithm in Table 5.1 resembles the multiblock PLS model proposed by MacGregor *et al.* [20] and Qin *et al.* [95]. In their work, a NIPALS algorithm has been applied to obtain $\mathbf{w}_{(j)}$ and $\boldsymbol{\beta}$, this algorithm is shown in Table 5.3. In [95], it was proven that $\begin{bmatrix} \boldsymbol{\beta}(0)\mathbf{w}_{(0)} \\ \vdots \\ \boldsymbol{\beta}(s-2)\mathbf{w}_{(s-2)} \\ \boldsymbol{\beta}(s-1)\mathbf{w}_{(s-1)} \end{bmatrix} \in \mathbb{R}^{sm}$ equals to $\mathbf{w}_{PLS}$, which is the eigenvector of $\mathbf{Y}_g\mathbf{\Theta}^T\mathbf{\Theta}\mathbf{Y}_g^T$ corresponding to $\lambda_{max}(\mathbf{Y}_g\mathbf{\Theta}^T\mathbf{\Theta}\mathbf{Y}_g^T)$. Furthermore, from Table 5.3, we can observe that for the ith iteration

$$\begin{aligned}
\boldsymbol{\beta}^i \propto \mathbf{t}_g^{i-1}(\mathbf{u}^i)^T &\propto \begin{bmatrix} \left(\mathbf{w}_{(0)}^i\right)^T\mathbf{Y}_{(0)} \\ \vdots \\ \left(\mathbf{w}_{(s-2)}^i\right)^T\mathbf{Y}_{(s-2)} \\ \left(\mathbf{w}_{(s-1)}^i\right)^T\mathbf{Y}_{(s-1)} \end{bmatrix} \mathbf{\Theta}^T\mathbf{q}^{i-1} \propto \begin{bmatrix} \left(\mathbf{w}_{(0)}^i\right)^T\mathbf{Y}_{(0)} \\ \vdots \\ \left(\mathbf{w}_{(s-2)}^i\right)^T\mathbf{Y}_{(s-2)} \\ \left(\mathbf{w}_{(s-1)}^i\right)^T\mathbf{Y}_{(s-1)} \end{bmatrix} \mathbf{\Theta}^T\mathbf{\Theta}\mathbf{t}^{i-1} \\
&\propto \begin{bmatrix} \left(\mathbf{w}_{(0)}^i\right)^T\mathbf{Y}_{(0)} \\ \vdots \\ \left(\mathbf{w}_{(s-2)}^i\right)^T\mathbf{Y}_{(s-2)} \\ \left(\mathbf{w}_{(s-1)}^i\right)^T\mathbf{Y}_{(s-1)} \end{bmatrix} \mathbf{\Theta}^T\mathbf{\Theta}\left(\mathbf{t}_g^{i-1}\right)\boldsymbol{\beta}^{i-1} \\
&= \begin{bmatrix} \left(\mathbf{w}_{(0)}^i\right)^T\mathbf{Y}_{(0)} \\ \vdots \\ \left(\mathbf{w}_{(s-2)}^i\right)^T\mathbf{Y}_{(s-2)} \\ \left(\mathbf{w}_{(s-1)}^i\right)^T\mathbf{Y}_{(s-1)} \end{bmatrix} \mathbf{\Theta}^T\mathbf{\Theta}\left[\ \mathbf{Y}_{(0)}^T\mathbf{w}_{(0)}^{i-1}, \ \cdots, \ \mathbf{Y}_{(s-2)}^T\mathbf{w}_{(s-2)}^{s-1}, \ \mathbf{Y}_{(s-1)}^T\mathbf{w}_{(s-1)}^{i-1}\ \right]\boldsymbol{\beta}^{i-1}
\end{aligned} \tag{5.9}$$

and

$$\mathbf{w}_{(j)}^i \propto \mathbf{Y}_{(j)}\mathbf{\Theta}\mathbf{\Theta}^T\mathbf{t}_G^{i-1}\boldsymbol{\beta}^{i-1} \propto \mathbf{Y}_{(j)}\mathbf{\Theta}\mathbf{\Theta}^T\left[\mathbf{Y}_{(0)}^T\mathbf{w}_{(0)}^{i-1}, ..., \mathbf{Y}_{(s-1)}^T\mathbf{w}_{(s-1)}^{i-1}\right]\boldsymbol{\beta}^{i-1} \tag{5.10}$$

Comparing Eq. (5.3) with Eqs. (5.9) and (5.10), it can be seen that when the NIPALS method converges, it will obtain the same $\boldsymbol{\beta}$ and $\mathbf{w}_{(j)}$ as the original algorithm. Similar to that developed in PLS model, Table 5.3 could be regarded as the power iteration method for resolving eigenvalue decomposition problem in Table 5.1.

Based on Table 5.3, the algorithm in Table 5.2 can be also realized using the NIPALS method. Table 5.4 shows the algorithm, where we can see that appropriate modifications are made in Steps (2)-(4) compared against Table 5.3.

Table 5.4: NIPALS algorithm for the IDPLS method

1. Initialize $\mathbf{u} = \mathbf{\Theta}(1,:)$, and randomly select $\boldsymbol{\beta}$
2. $\mathbf{w}_{(j)} = \mathbf{Y}_{(j)}\mathbf{u}^T$, $j = 0, ..., s-1$
3. $\mathbf{w} = \sum_j \boldsymbol{\beta}_{(j)}\mathbf{w}_{(j)}$, $\mathbf{w} = \mathbf{w}/\|\mathbf{w}\|_{\mathrm{E}}$, $j = 0, ..., s-1$
4. $\mathbf{t}_{(j)} = \mathbf{w}^T\mathbf{Y}_{(j)}$, $j = 0, ..., s-1$
5. $\mathbf{t}_G = \left[\mathbf{t}_{(0)}^T, \cdots, \mathbf{t}_{(s-2)}^T, \mathbf{t}_{(s-1)}^T\right]^T$
6. $\boldsymbol{\beta} = \mathbf{t}_G\mathbf{u}^T / \left\|\mathbf{t}_G\mathbf{u}^T\right\|_{\mathrm{E}}$
7. $\mathbf{t}_g = \boldsymbol{\beta}^T\mathbf{t}_G$
8. $\mathbf{q} = \mathbf{\Theta}\mathbf{t}_g^T / \mathbf{t}_g\mathbf{t}_g^T$
9. $\mathbf{u} = \mathbf{q}^T\mathbf{\Theta} / \mathbf{q}^T\mathbf{q}$
10. Return to Step 2 until the convergence is reached.

Note that in Table 5.4, for the ith iteration

$$\begin{aligned}
\mathbf{w}^i &\propto \left(\boldsymbol{\beta}^{i-1} \otimes \mathbf{I}_m\right) \begin{bmatrix} \mathbf{w}_{(0)}^i \\ \vdots \\ \mathbf{w}_{(s-1)}^i \end{bmatrix} \propto \left(\boldsymbol{\beta}^{i-1} \otimes \mathbf{I}_m\right) \mathbf{Y}_g \left(\mathbf{u}^{i-1}\right)^T \\
&\propto \left(\boldsymbol{\beta}^{i-1} \otimes \mathbf{I}_m\right)^T \mathbf{Y}_g \mathbf{\Theta}^T \mathbf{\Theta} \mathbf{t}_g^{i-1} \boldsymbol{\beta}^{i-1} \\
&= \left(\boldsymbol{\beta}^{i-1} \otimes \mathbf{I}_m\right)^T \mathbf{Y}_g \mathbf{\Theta}^T \mathbf{\Theta} \mathbf{Y}_g^T \left(\mathbf{I}_m \otimes \mathbf{w}^{i-1}\right) \boldsymbol{\beta}^{i-1} \\
&= \left(\boldsymbol{\beta}^{i-1} \otimes \mathbf{I}_m\right)^T \mathbf{Y}_g \mathbf{\Theta}^T \mathbf{\Theta} \mathbf{Y}_g^T \left(\boldsymbol{\beta}^{i-1} \otimes \mathbf{I}_m\right) \mathbf{w}^{i-1}
\end{aligned} \tag{5.11}$$

and

$$\begin{aligned}
\boldsymbol{\beta}^i &\propto \mathbf{t}_g^i \left(\mathbf{u}^{i-1}\right)^T \propto \left(\mathbf{I}_s \otimes \mathbf{w}^i\right)^T \mathbf{Y}_g \mathbf{\Theta}^T \mathbf{\Theta} \left(\mathbf{t}_g^i\right)^T \boldsymbol{\beta}^{i-1} \\
&= \left(\mathbf{I}_s \otimes \mathbf{w}^i\right)^T \mathbf{Y}_g \mathbf{\Theta}^T \mathbf{\Theta} \mathbf{Y}_g^T \left(\mathbf{I}_s \otimes \mathbf{w}^{i-1}\right) \boldsymbol{\beta}^{i-1}
\end{aligned} \tag{5.12}$$

Comparing Eqs. (5.11) and (5.12) with Eqs. (5.7), it can be found that after reaching convergence, the method in Table 5.4 is equivalent to the original one in Table 5.2. Note that convergence of this algorithm can be guaranteed similarly to that given by the NIPALS algorithm for PLS models [92].

5.2.3 Deflations and the complete DPLS model

The above sections primarily emphasize the optimization problem included in DPLS methods. The $\mathbf{w}_{(j)}$ and $\boldsymbol{\beta}$ obtained using algorithms either in Tables 5.1 and 5.2 or Tables 5.3 and 5.4 should be subsequently used to derive the two scores: $\mathbf{t}_{(j)} \in \mathbb{R}^{1\times N} = \mathbf{w}_{(j)}^T \mathbf{Y}_{(j)}$ and $\mathbf{t}_g \in \mathbb{R}^{1\times N} = \sum_{j=0}^{s-1} \boldsymbol{\beta}(j)\,\mathbf{t}_{(j)}$. Next, the deflation procedure is implemented [30, 80].

At first, two loading vectors are configured as $\mathbf{p}_{(j)} = \mathbf{Y}_{(j)}\mathbf{t}_{(j)}^T \Big/ \mathbf{t}_{(j)}\mathbf{t}_{(j)}^T$ and $\mathbf{q} = \boldsymbol{\Theta}\mathbf{t}_g^T / \mathbf{t}_g \mathbf{t}_g^T$, next $\mathbf{Y}$ and $\boldsymbol{\Theta}$ are deflated by

$$\begin{aligned}\bar{\mathbf{Y}}_{(j)} &= \mathbf{Y}_{(j)} - \mathbf{p}_{(j)}\mathbf{t}_{(j)} \\ \bar{\mathbf{Y}} &= [\bar{\mathbf{Y}}_{(s-1)}^T, \bar{\mathbf{Y}}_{(s-2)}^T, \cdots, \bar{\mathbf{Y}}_{(0)}^T]^T \\ \bar{\boldsymbol{\Theta}} &= \boldsymbol{\Theta} - \mathbf{q}\mathbf{t}_g = \boldsymbol{\Theta} - \mathbf{q}\sum_{j=0}^{s-1} \boldsymbol{\beta}(j)\mathbf{t}_{(j)}\end{aligned} \tag{5.13}$$

Then, let $\bar{\mathbf{Y}}$ and $\bar{\boldsymbol{\Theta}}$ replace $\mathbf{Y}$ and $\boldsymbol{\Theta}$, repeat the methods in Tables 5.1 and 5.2 or Tables 5.3 and 5.4 until some stopping criterion is reached. Suppose κ deflations have been completed, then,

$$\mathrm{B} = [\boldsymbol{\beta}_1, \boldsymbol{\beta}_2, \ldots, \boldsymbol{\beta}_\kappa] \in \mathbb{R}^{s\times\kappa} \tag{5.14a}$$

$$\mathbf{T}_g = \left[\mathbf{t}_{g1}^T, \mathbf{t}_{g2}^T, \cdots, \mathbf{t}_{g\kappa}^T\right]^T \in \mathbb{R}^{\kappa\times N} \tag{5.14b}$$

$$\mathbf{T}_{(j)} = \left[\mathbf{t}_{(j),1}^T, \mathbf{t}_{(j),2}^T, \cdots, \mathbf{t}_{(j),\kappa}^T\right]^T \in \mathbb{R}^{\kappa\times N} \tag{5.14c}$$

$$\mathbf{Q} = [\mathbf{q}_1, \mathbf{q}_2 ..., \mathbf{q}_\kappa] \in \mathbb{R}^{l\times\kappa} \tag{5.14d}$$

$$\mathbf{P}_{(j)} = \left[\mathbf{p}_{(j)1}, \mathbf{p}_{(j)2}, ..., \mathbf{p}_{(j)}\kappa\right] \in \mathbb{R}^{m\times\kappa} \tag{5.14e}$$

Here, $\mathbf{Q}$ matrix has a similar definition with that adopted in PLS model. Note that for the IDPLS method, as the common $\mathbf{w}_{(j)} = \mathbf{w}$ is assumed, thus, $\mathbf{W} = [\mathbf{w}_1, \mathbf{w}_2, ..., \mathbf{w}_\kappa]$ can be finally obtained. While in the DDPLS method, $\mathbf{W}_{(j)} = \left[\mathbf{w}_{(j)1}, \mathbf{w}_{(j)2}, ..., \mathbf{w}_{(j)\kappa}\right]$ $\forall j$ are obtained. Similar to the PLS model, the weighting matrix $\mathbf{R}_{(j)}$ can be defined for both methods as

$$\mathbf{R}_{(j)} = \begin{cases} \mathbf{W}\left(\mathbf{P}_{(j)}^T\mathbf{W}\right)^{-1} \in \mathbb{R}^{m\times\kappa} & \text{for IDPLS} \\ \mathbf{W}_{(j)}\left(\mathbf{P}_{(j)}^T\mathbf{W}_{(j)}\right)^{-1} \in \mathbb{R}^{m\times\kappa} & \text{for DDPLS}\end{cases} \tag{5.15}$$

which gives $\mathbf{T}_{(j)} = \mathbf{R}_{(j)}^T\mathbf{Y}_{(j)}$ $j = 0, ..., s-1$. Let $\mathcal{B}_j = \begin{bmatrix} \boldsymbol{\beta}_1(j) & & \\ & \ddots & \\ & & \boldsymbol{\beta}_\kappa(j)\end{bmatrix} \in \mathbb{R}^{\kappa\times\kappa}$ with $j = 1, ..., s$.

Finally, the model given by $\mathbf{Y}_{(j)}$ and $\boldsymbol{\Theta}$ for the two DPLS methods is

$$\mathbf{Y}_{(j)} = \mathbf{P}_{(j)}\mathbf{T}_{(j)} + \tilde{\mathbf{Y}}_{(j)} \tag{5.16a}$$

$$\boldsymbol{\Theta} = \mathbf{Q}\sum_{j=1}^{s} \mathcal{B}_j \mathbf{T}_{(j-1)} + \tilde{\boldsymbol{\Theta}} \tag{5.16b}$$

where $\tilde{\mathbf{Y}}_{(j)} \in \mathbb{R}^{m \times N}$ and $\tilde{\boldsymbol{\Theta}} \in \mathbb{R}^{l \times N}$ denote unmodelled parts of $\mathbf{Y}_{(j)}$ and $\boldsymbol{\Theta}$ by DPLS. Online implementation of the DPLS model is given, at time k, as:

$$\begin{aligned} \mathrm{t}_{(j)} &= \mathbf{R}_{(j)}^{T}\mathbf{y}(k-j) \in \mathbb{R}^{\kappa} \\ \hat{\boldsymbol{\theta}}(k) &= \mathbf{Q}\sum_{j=1}^{s} \mathcal{B}_j \mathrm{t}_{(j)} = \mathbf{Q}\mathrm{t}_g \end{aligned} \tag{5.17}$$

where $\mathbf{y}(k-j)$ denotes the measurements at time instance $k-j$, with $j = 0, ..., s-1$.

Two parameters, s and κ, should be specified before using the two DPLS methods. Li *et al.* [12] have proposed a cross-validation-based method, where the pair of s and κ that give the smallest predicted residual sum of squares (PRESS) value is used. Zhang *et al.* [89] have proposed a mixture of cross-validation and the Akaike Information Criterion (AIC) method in order to give a trade-off between accuracy and complexity. Note that κ is consist with $\bar{m}$ in PCA model (Eqs. (4.14) and (2.1)) and κ in PLS model (Section 4.2.3).

5.3 EDD-based performance evaluation

5.3.1 KPI-based monitoring using DPLS models

Due to measurement issues with key components of the KPI, they may not be available online as often or as accurately as required. Thus, online monitoring of KPIs needs to be performed using PLS- or DPLS-models to provide the missing values. PLS-based KPI monitoring approaches [71] follow the idea that given a new coming $\mathbf{y}_{new}$, there will be a predicted KPI $\hat{\boldsymbol{\theta}}_{new}$ using the PLS method. Then, the T^2 statistic can be obtained from $\hat{\boldsymbol{\theta}}_{new}$ as $J_\theta = \hat{\boldsymbol{\theta}}_{new}^T \Sigma_{\hat{\theta}}^{-1} \hat{\boldsymbol{\theta}}_{new}$. For dynamic processes, it is less reasonable to directly adopt the same idea since $\hat{\boldsymbol{\theta}}$ is autocorrelated. An intuitive approach is to fit a VAR model [96] to $\hat{\boldsymbol{\theta}}$ as

$$\hat{\boldsymbol{\theta}}(k) = \mathbb{A}_1\hat{\boldsymbol{\theta}}(k-1) + \cdots + \mathbb{A}_{\tau_1}\hat{\boldsymbol{\theta}}(k-\tau_1) + \mathbf{v}_\theta(k) \tag{5.18}$$

As it is assumed that $\mathbf{v}_\theta \sim \mathcal{N}_l(0, \Sigma_{v_\theta})$, $J_\theta = \mathbf{v}_\theta(k)^T \Sigma_{v\theta}^{-1} \mathbf{v}_\theta(k) \sim \chi_l^2$ can be designed for PM-FD of KPI. Considering that $\hat{\boldsymbol{\theta}} = \mathbf{Q}\mathrm{t}_g$, an alternative method is to directly apply the model given by Eq. (5.18) to $\mathbf{t}_g$, that is,

$$\mathrm{t}_g(k) = \mathbb{C}_1\mathrm{t}_g(k-1) + \cdots + \mathbb{C}_{\tau 2}\mathrm{t}_g(k-\tau_2) + \mathbf{v}_g(k) \tag{5.19}$$

Likewise, $J_\theta = \mathbf{v}_g(k)^T \Sigma_{v_g}^{-1} \mathbf{v}_g(k) \sim \chi_\kappa^2$ can be developed. Let t be the shorthand for $\mathrm{t}_{(0)}$, then $\mathrm{t} = \mathbf{R}_{(0)}^T \mathbf{y}$. In [12], a direct VAR model on t is given as

$$\mathrm{t}(k) = \mathbb{D}_1\mathrm{t}(k-1) + \cdots + \mathbb{D}_\tau\mathrm{t}(k-\tau) + \mathbf{v}_t(k) \tag{5.20}$$

which gives $J_\theta = \mathbf{v}_t(k)^T \Sigma_{\mathbf{v}_t}^{-1} \mathbf{v}_t(k) \sim \chi_\kappa^2$. In Eqs. (5.18)-(5.20), parameters $\{\mathbb{A}_1, \cdots, \mathbb{A}_{\tau_1}; \mathbb{C}_1, \cdots, \mathbb{C}_{\tau_2}; \mathbb{D}_1, \cdots, \mathbb{D}_\tau\}$ can be estimated by LS-based methods, τ, τ_1 and τ_2 can be determined using AIC [96]. The thresholds are calculated by $J_{th,\theta} = \chi_{l,\alpha}^2$ for Eq. (5.20) and $J_{th,\theta} = \chi_{\kappa,\alpha}^2$ for Eqs. (5.19) and (5.20). Compared with the first two methods, the third one is more straightforward to use. Thus, it is combined with the two DPLS methods to create a KPI-based PM-FD method.

5.3.2 Performance evaluation with respect to EDD

In this section, the EDD index is applied to assess the performance of two DPLS-based KPI monitoring approaches. Considering a constant additive fault expressed as $\mathbf{y}_f = \mathbf{y} + \mathbf{f}$, when it occurs at time k_s, $\mathbf{v}_t$ is

$$\begin{aligned}\mathbf{v}_t(k_s) &= \mathbf{R}^T(\mathbf{y}^*(k_s) + f) - \mathbf{D}_1\mathbf{R}^T\mathbf{y}(k_s-1) - \cdots - \mathbf{D}_\tau\mathbf{R}^T\mathbf{y}(k_s-\tau) \\ &\sim \mathcal{N}_\kappa\left(\mathbf{R}^T\mathbf{f}, \Sigma_{\mathbf{v}_t}\right)\end{aligned} \tag{5.21}$$

where $\mathbf{y}^*$ is the fault-free data, and $\mathbf{R}$ is the shorthand for $\mathbf{R}_{(0)}$ and has a similar definition with that proposed in Chapter 4 for PLS model. Thus, $\mathrm{FDR}(k_s)$ is calculated as

$$\mathrm{FDR}(k_s) = 1 - F_{\chi_\kappa^2(\delta)}(J_{th,\theta}) \tag{5.22}$$

where $\delta = \mathbf{f}^T\mathbf{R}\Sigma_{\mathbf{v}_t}^{-1}\mathbf{R}^T\mathbf{f}$. Iteratively, for $k = k_s + 1$ to $k_s + \tau$, $\mathbf{v}_t$ is calculated as

$$\begin{aligned}\mathbf{v}_t(k_s+i) &= \mathbf{R}^T(\mathbf{y}^*(k_s+i)+\mathbf{f}) - \mathbb{D}_1\mathbf{R}^T(\mathbf{y}^*(k_s+i-1)+\mathbf{f}) \\ &\quad - \mathbb{D}_i\mathbf{R}^T(\mathbf{y}^*(k_s)+\mathbf{f})\cdots - \mathbb{D}_\tau\mathbf{R}^T\mathbf{y}(k_s-\tau+i) \\ &\sim \mathcal{N}_\kappa\left(\left(\mathbf{R}^T - \sum_{j=1}^{i}\mathbb{D}_j\mathbf{R}^T\right)\mathbf{f}, \Sigma_{\mathbf{v}_t}\right)\end{aligned} \tag{5.23}$$

Then, FDR(k_s+i) is likewise obtained according to Eq. (5.22). Note that $\forall i \geq \tau$, $\mathbf{v}_t \sim \mathcal{N}_\kappa\left(\left(\mathbf{R}^T - \sum_{j=1}^{\tau}\mathbb{D}_j\mathbf{R}^T\right)\mathbf{f}, \Sigma_{\mathbf{v}_t}\right)$ is satisfied, and thus, they give the constant FDR value. Substituting all FDR values into Eq. (2.13), based on Theorem 2.2, it satisfies that $\sum_{k=0}^{\infty}\text{prob}(\mathcal{J}=k) = 1$.

5.4 Simulation results

In order to verify the theoretical results and explore some of the implications of the above results, a numerical case will be performed. The simulated example without faults is

$$\begin{cases} \mathrm{t}(k) = \mathbf{\Phi}_1\mathrm{t}(k-1) - \mathbf{\Phi}_2\mathrm{t}(k-2) + \mathrm{t}^*(k) \\ \mathbf{y}(k) = \mathbf{P}\mathrm{t}(k) + \boldsymbol{\nu}(k) \\ \boldsymbol{\theta}(k) = \mathbf{\Psi}_1\mathbf{y}(k) + \mathbf{\Psi}_2\mathbf{y}(k-1) + \boldsymbol{\zeta}(k) \end{cases} \tag{5.24}$$

where $\mathrm{t}^*(k) \sim \mathcal{N}_3(0, \mathbf{I}_3)$, $\boldsymbol{\nu}(k) \sim \mathcal{N}_5(0, 0.01\mathbf{I}_5)$, $\boldsymbol{\zeta}_k \sim \mathcal{N}_2(0, 0.02\mathbf{I}_2)$, and

$$\mathbf{\Phi}_1 = \begin{bmatrix} 0.4389 & 0.1210 & -0.0862 \\ -0.2966 & -0.0550 & 0.2274 \\ 0.4538 & -0.6573 & 0.4239 \end{bmatrix}, \mathbf{\Phi}_2 = \begin{bmatrix} -0.2998 & -0.1905 & -0.2669 \\ -0.0204 & -0.1585 & -0.2950 \\ 0.1461 & -0.0755 & 0.3749 \end{bmatrix}$$

$$\mathbf{P} = \begin{bmatrix} 0.5586 & 0.2042 & 0.6370 \\ 0.2007 & 0.0492 & 0.4429 \\ 0.0874 & 0.6062 & 0.0664 \\ 0.9332 & 0.5463 & 0.3743 \\ 0.2594 & 0.0958 & 0.2491 \end{bmatrix}, \mathbf{\Psi}_1 = \begin{bmatrix} 0.9249 & 0.4350 \\ 0.6295 & 0.9811 \\ 0.8743 & 0.0960 \\ 0.6417 & 0.5275 \\ 0.7984 & 0.5456 \end{bmatrix}^T, \mathbf{\Psi}_2 = \begin{bmatrix} 1.7198 & -0.3715 \\ 0.5835 & 1.5011 \\ 1.4236 & 1.3226 \\ 0.4963 & -1.4145 \\ -2.5717 & 1.0696 \end{bmatrix}^T$$

Comparison of the two DPLS models

Let $n = 1000$. Using Eq. (5.24), generate 1000 samples of $\mathbf{y}_{obs}$ and $\boldsymbol{\theta}_{obs}$. After the normalization, $\mathbf{y}$ and $\boldsymbol{\theta}$ are available for building the two DPLS

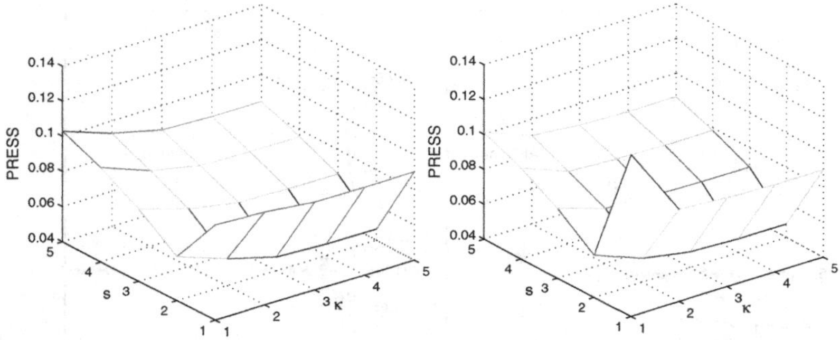

Figure 5.1: Cross-validation results in the numerical example: (left) the IDPLS method; (right) the DDPLS method.

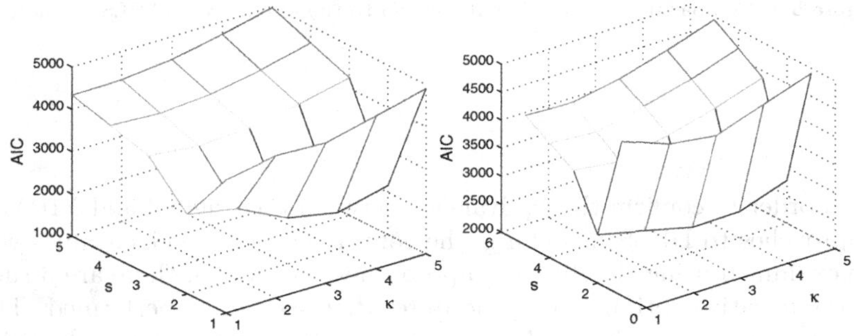

Figure 5.2: The mixture of AIC and cross-validation result in the numerical example: (left) the IDPLS method; (right) the DDPLS method.

models. The first step is to select κ and s. Applying the cross-validation-based method [12] leads to the results shown in Figure 5.1. We can see that $\kappa = 3$ or $4, s = 2$ is determined for this case. Furthermore, the mixture of AIC and cross-validation-based method developed in [89] gives the results shown in Figure 5.2. It can be found that the two methods agree well with each other for this example. The only difference is that the second method which takes into account the modeling complexity gives $\kappa = 3, s = 2$.

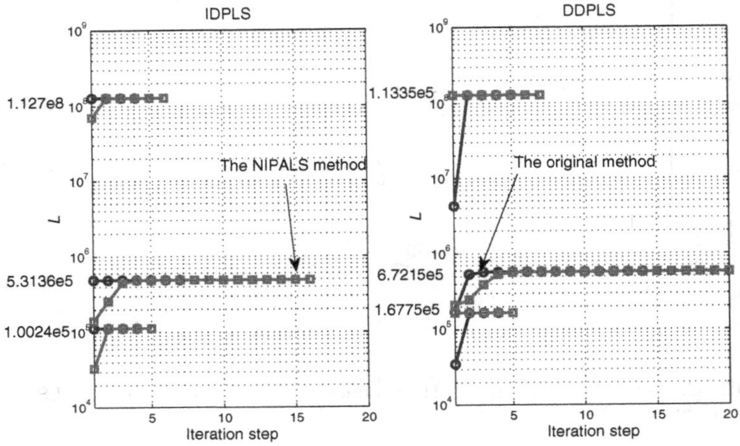

Figure 5.3: Comparison of the original method to the alterative NIAPLS method

In order to confirm the equivalence between the original and NIPALS approaches to DPLS modelling, the convergence of the eigenvalues will be examined using the different approaches. For $\kappa = 3$, there are 3 runs of the iterative optimization procedures that need to be performed. The results are shown in Figure 5.3, where it can be seen that for both DPLS methods, the two type of approaches converge to the same maximum $\mathcal{L}_{max}$. Comparing the two NIPALS methods, since the one in Table 5.4 has more constraints, it can be seen that the IDPLS method needs a few more steps to converge. Furthermore, it should be noted that the two DPLS methods can converge to different $\mathcal{L}_{max}$ values. Since DDPLS involves a constraint with identical $\mathbf{w}_{(j)}$, it gives a smaller $\mathcal{L}_{max}$ in the three runs of optimisation. After obtaining $\mathbf{T}_{(0)} \in \mathbb{R}^{3\times 800}$, the VAR model is developed for both methods. The parameter $\tau = 2$ is determined using the AIC index as shown in Figure 5.4. Then the residual vector, $\mathbf{v}_t$, is obtained using Eq. (5.20). Figure 5.5 shows the autocorrelation test for t and $\mathbf{v}_t$. We can see that the residual variables are static and, thus, more appropriate for formulating a detection test statistic.

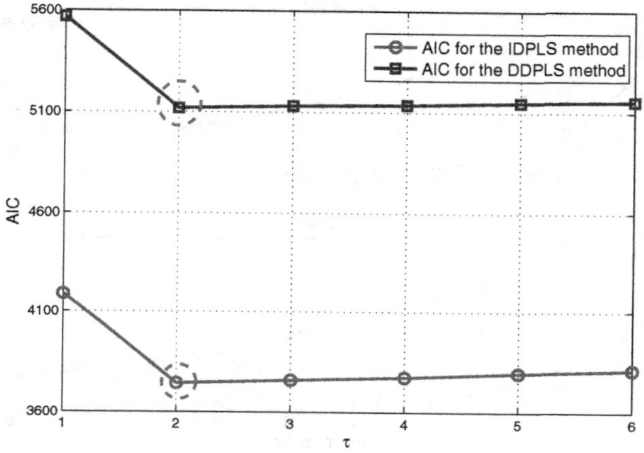

Figure 5.4: AIC results of the VAR model in the numerical example

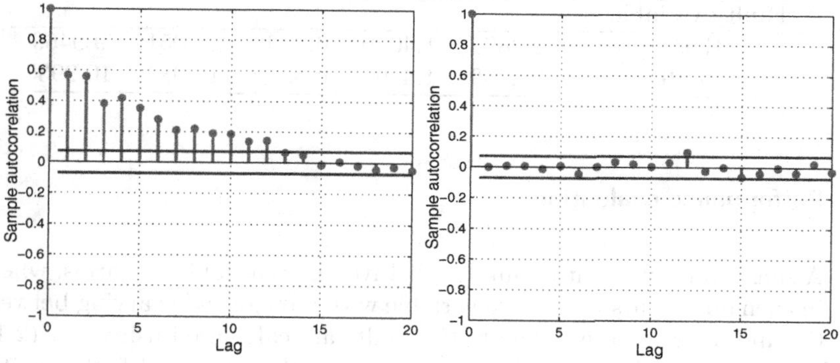

Figure 5.5: Comparison of autocorrelation coefficients before (left) and after performing VAR model on **t** (right)

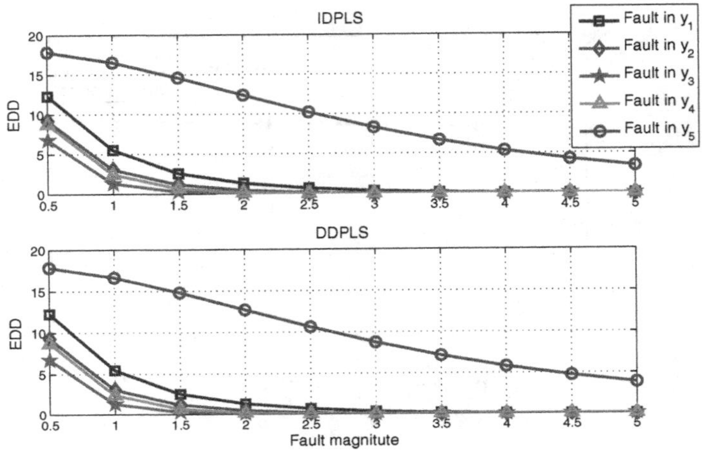

Figure 5.6: Comparison of EDD in the numerical example

Table 5.5: Comparison of the average EDD given by two DPLS methods

Faulty variable #	1	2	3	4	5
IDPLS	2.2898	1.4029	0.8408	1.2051	9.9469
DDPLS	2.2739	1.4196	0.8389	1.1987	10.2935

Performance evaluation

A single variable fault is simulated. Five different fault scenarios, wherein each one occurs in a single variable with a magnitude varying between 0.5 and 5, are created. The EDD results are calculated using Eq. (2.15) and shown in Figure 5.6. It can be observed that for all faults, as the magnitude increases, EDD likewise decreases to 0. For different faults, EDD performance is different. In general, the two DPLS methods perform similarly in all fault scenarios. Taking the average EDD for each fault, the results are shown in Table 5.5. It can be seen that when faults occur in variable 1 and 5, IDPLS behaves better, while for other cases, DDPLS performs better.

5.5 Conclusions

To address the KPI-based FD problem for dynamic processes that work in the steady state, DPLS models have been extensively used both in research and practice. This chapter examined two popular DPLS methods: the DDPLS method, which computes the optimal decomposition based on individual direction vectors, and the IDPLS method, which computes the optimal decomposition based on the assumption that the direction vectors for all eigenvectors is the same. Furthermore, a NIPALS algorithm has been proposed for these two methods in order to avoid performing a matrix decomposition. It was shown that the NIPALS approach is identical to the original formulation. Finally, the two methods were compared in terms of the their fault detection ability with respect to the EDD index.

6 KPI-based PM-FD methods for dynamic processes

As discussed in Chapter 1, the most commonly applied representation of the dynamic process model is the state-space form. This chapter introduces KPI-based PM-FD methods that based on the state-space model to represent the relation between KPI and process variables. Two methods will be presented using the idea of data-driven realisation of parity space.

6.1 Background

Based on Eq. (2.6), Ding [2, 6] has proposed that, theoretically, the parity space equals the kernel representation of a dynamic system, such that the data-driven form of kernel representation is

$$\forall \mathbf{u}_s(k), \mathbf{x}(0), \mathbf{r}(k) = \mathcal{K}_s \begin{bmatrix} \mathbf{u}_s(k) \\ \mathbf{y}_s(k) \end{bmatrix} \tag{6.1}$$

where $\mathbf{u}_s$ and $\mathbf{y}_s$ denote the process input and output data in the time interval $[k-s, k]$, which is consistent with that used in Section 5.1 and $\mathbf{r}(k)$ represents the noise information. One way to solve $\mathcal{K}_s$ is to utilize the historical process data [6, 33]. First, rewrite model (2.6) into the form

$$\begin{bmatrix} \mathbf{U}_{k,s} \\ \mathbf{Y}_{k,s} \end{bmatrix} = \boldsymbol{\Psi}_s \begin{bmatrix} \mathbf{U}_{k,s} \\ \mathbf{X}_{k-s} \end{bmatrix} + \begin{bmatrix} \mathbf{0} \\ \mathbf{H}_{w,s}\mathbf{W}_{k,s} + \mathbf{V}_{k,s} \end{bmatrix} \tag{6.2}$$

where $\Psi_s = \begin{bmatrix} \mathbf{I}_{(s+1)l} & \mathbf{0} \\ \mathbf{H}_{u,s} & \mathbf{\Gamma}_s \end{bmatrix}$, $\mathbf{\Gamma}_s = \begin{bmatrix} \mathbf{C} \\ \mathbf{CA} \\ \vdots \\ \mathbf{CA}^s \end{bmatrix}$,

$\mathbf{U}_{k,s} = \begin{bmatrix} \mathbf{u}(k-s) & \dots & \mathbf{u}(k-s+N-1) \\ \vdots & \ddots & \vdots \\ \mathbf{u}(k) & \dots & \mathbf{u}(k+N-1) \end{bmatrix}$,

$\mathbf{H}_{u,s} = \begin{bmatrix} \mathbf{D} & \mathbf{0} & & \\ \mathbf{CB} & \ddots & & \\ \vdots & \ddots & \ddots & \mathbf{0} \\ \mathbf{CA}^{s-1}\mathbf{B} & \dots & \mathbf{CB} & \mathbf{D} \end{bmatrix}$, $\mathbf{X}_{k-s} = \begin{bmatrix} \mathbf{x}(k-s) & \dots & \mathbf{x}(k-s-N+1) \end{bmatrix}$, $\mathbf{Y}_{k,s}$ and $\mathbf{H}_{w,s}$ have the same structure with $\mathbf{U}_{k,s}$. Let $\mathcal{K}_s = (\Psi_s^{\perp})^T$, which leads to

$$\mathcal{K}_s \begin{bmatrix} \mathbf{U}_{k,s} \\ \mathbf{Y}_{k,s} \end{bmatrix} = \mathcal{K}_s \begin{bmatrix} \mathbf{0} \\ \mathbf{H}_{w,s}\mathbf{W}_{k,s} + \mathbf{V}_{k,s} \end{bmatrix} \tag{6.3}$$

Let $\mathcal{Z}_p = \mathcal{Z}_{k-s-1,s} = \begin{bmatrix} \mathbf{U}_{k-s-1,s} \\ \mathbf{Y}_{k-s-1,s} \end{bmatrix}$, and performing a QR decomposition gives [6, 98]

$$\begin{bmatrix} \mathcal{Z}_p \\ \mathbf{U}_{k,s} \\ \mathbf{Y}_{k,s} \end{bmatrix} = \begin{bmatrix} \mathbf{R}_{11} & \mathbf{0} & \mathbf{0} \\ \mathbf{R}_{21} & \mathbf{R}_{22} & \mathbf{0} \\ \mathbf{R}_{31} & \mathbf{R}_{32} & \mathbf{R}_{33} \end{bmatrix} \begin{bmatrix} \mathbf{Q}_1 \\ \mathbf{Q}_2 \\ \mathbf{Q}_3 \end{bmatrix} \tag{6.4}$$

which leads to

$$\begin{bmatrix} \mathbf{U}_{k,s} \\ \mathbf{Y}_{k,s} \end{bmatrix} = \begin{bmatrix} \mathbf{R}_{21} & \mathbf{R}_{22} \\ \mathbf{R}_{31} & \mathbf{R}_{32} \end{bmatrix} \begin{bmatrix} \mathbf{Q}_1 \\ \mathbf{Q}_2 \end{bmatrix} + \begin{bmatrix} 0 \\ \mathbf{R}_{33}\mathbf{Q}_3 \end{bmatrix} \tag{6.5}$$

and

$$\mathbf{H}_{w,s}\mathbf{W}_{k,s} + \mathbf{V}_{k,s} = \mathbf{R}_{33}\mathbf{Q}_3 \tag{6.6}$$

Further take an SVD on

$$\begin{bmatrix} \mathbf{R}_{21} & \mathbf{R}_{22} \\ \mathbf{R}_{31} & \mathbf{R}_{32} \end{bmatrix} = \begin{bmatrix} \mathbf{P}_R & \tilde{\mathbf{P}}_R \end{bmatrix} \begin{bmatrix} \Lambda_R & \\ & \mathbf{0} \end{bmatrix} \mathbf{V}_R^T \tag{6.7}$$

Thus, let $\mathcal{K}_s \in \mathbb{R}^{1\times(s+1)(m+l)}$ be any row of $\tilde{\mathbf{P}}_R^T$, then

$$\mathcal{K}_s \begin{bmatrix} \mathbf{R}_{21} & \mathbf{R}_{22} \\ \mathbf{R}_{31} & \mathbf{R}_{32} \end{bmatrix} \begin{bmatrix} \mathbf{Q}_1 \\ \mathbf{Q}_2 \end{bmatrix} = 0 \tag{6.8}$$

Furthermore, it gives

$$\mathcal{K}_s \begin{bmatrix} \mathbf{U}_{k,s} \\ \mathbf{Y}_{k,s} \end{bmatrix} = \mathcal{K}_{s,y} \left(\mathbf{H}_{w,s} \mathbf{W}_{k,s} + \mathbf{V}_{k,s} \right) = \mathcal{K}_{s,y} \mathbf{R}_{33} \mathbf{Q}_3 \tag{6.9}$$

where $\mathcal{K}_{s,y}$ equals the lower part of $\mathcal{K}_s$. In [113], a robust method was developed to select it. Extension of it to time-varying processes can be found in [124], where a recursive design is used. Regarding to each subsystem, the data-driven kernel representations of the system can be obtained based on the historical data.

6.1.1 Parity-space-based fault detection

Based on the data-driven realization for the kernel representation of a dynamic system, the PS-based PM-FD method can be developed. Let $\mathbf{\Psi}_s^{\perp} = \tilde{\mathbf{P}}_R \in \mathbb{R}^{(s+1)(m+l)\times((s+1)l-n)}$, a residual generator is built as

$$\mathbf{r}(k) = \left(\mathbf{\Psi}_s^{\perp}\right)^T \begin{bmatrix} \mathbf{u}_s(k) \\ \mathbf{y}_s(k) \end{bmatrix} \in \mathbb{R}^{(s+1)l-n} \sim \mathcal{N}(0, \Sigma_{\mathbf{r}}) \tag{6.10}$$

where $\Sigma_{\mathbf{r}} = \frac{\left(\mathbf{\Psi}_s^{\perp}\right)^T \mathbf{R}_{33} \mathbf{R}_{33}^T \mathbf{\Psi}_s^{\perp}}{N-1}$. For the FD purpose, the T^2 statistic is given

$$T_{\mathbf{r}}^2 = \mathbf{r}^T \Sigma_{\mathbf{r}}^{-1} \mathbf{r} \sim \chi^2_{(s+1)l-n} \tag{6.11}$$

The corresponding threshold is determined by $J_{th,T_{\mathbf{r}}^2} = \chi^2_{(s+1)l-n,\alpha}$. Finally, the decision logic is

$$\begin{cases} T_{\mathbf{r}}^2 \leq J_{th,T_{\mathbf{r}}^2} \Rightarrow \text{fault} - \text{free} \\ T_{\mathbf{r}}^2 > J_{th,T_{\mathbf{r}}^2} \Rightarrow \text{A fault occurs} \end{cases} \tag{6.12}$$

It is be noted that for this method s should be selected much larger than the real system order n. This would increase the online computation costs and memory requirements since decision (6.12) is made based on $s+1$ consecutive measurements. As a result, it is needed to develop a method with recursive form.

6.1.2 Data-driven diagnostic observer

The interconnection between DO and parity space has been well studied in model-based FD methods. Using the parity vectors $\mathbf{\Psi}_s^{\perp}$ achieved by data-driven methods, we can also design the data-based diagnostic observer. Let

$$\begin{aligned} \boldsymbol{\alpha}_s &= \left[\boldsymbol{\alpha}_{s,0}, \boldsymbol{\alpha}_{s,1}, ..., \boldsymbol{\alpha}_{s,s}\right] \in \mathbb{R}^{1\times(s+1)m} \\ \boldsymbol{\beta}_s &= \left[\boldsymbol{\beta}_{s,0}, \boldsymbol{\beta}_{s,1}, ..., \boldsymbol{\beta}_{s,s}\right] \in \mathbb{R}^{1\times(s+1)l} \end{aligned} \tag{6.13}$$

and $[\boldsymbol{\beta}_s, \boldsymbol{\alpha}_s]$ is selected as any row of $\left(\mathbf{\Psi}_s^{\perp}\right)^T$. Since $\boldsymbol{\alpha}_s$ is a parity vector, it should be noted that $\boldsymbol{\beta}_s = -\alpha_s \mathbf{H}_{u,s}$. Based on α_s and $\boldsymbol{\beta}_s$, the DO could be designed as

$$\begin{aligned} \mathbf{x}_d\left(k+1\right) &= \mathbf{G}\mathbf{x}_d\left(k\right) + \mathbf{H}\mathbf{u}\left(k\right) + \mathbf{L}\mathbf{y}\left(k\right) \\ \mathbf{r}\left(k\right) &= \mathbf{g}\mathbf{y}\left(k\right) - \mathbf{c}\mathbf{x}_d\left(k\right) - \mathbf{d}\mathbf{u}\left(k\right) \end{aligned} \tag{6.14}$$

where $\mathbf{x}_d(k) = \mathbb{T}\mathbf{x}(k) \in \mathbb{R}^s$ and $\mathbb{T}$ is obtained as

$$\mathbb{T} = \begin{bmatrix} \alpha_{s,1} & \alpha_{s,2} & \cdots & \alpha_{s,s} \\ \alpha_{s,2} & \alpha_{s,3} & \cdots & \mathbf{0} \\ \vdots & \vdots & \vdots & \vdots \\ \alpha_{s,s} & \mathbf{0} & \cdots & \mathbf{0} \end{bmatrix} \begin{bmatrix} \mathbf{C} \\ \mathbf{CA} \\ \vdots \\ \mathbf{CA}^{s-1} \end{bmatrix} \in \mathbb{R}^{s\times n}$$

Let $\mathbf{G} = \begin{bmatrix} 0 & 0 & 0 & 0 & 0 \\ 1 & 0 & \cdots & 0 & 0 \\ \vdots & \ddots & \ddots & \vdots & \vdots \\ 0 & \cdots & 1 & 0 & 0 \\ 0 & \cdots & 0 & 1 & 0 \end{bmatrix} \in \mathbb{R}^{s\times s}$, which guarantees that the observer is stable. The remaining parameters could be obtained by solving the Luenberger equations:

$$\mathbb{T}\mathbf{A} - \mathbf{G}\mathbb{T} = \mathbf{L}\mathbf{C}, \mathbf{H} = \mathbb{T}\mathbf{B} - \mathbf{L}\mathbf{D}, \mathbf{g}\mathbf{C} = \mathbf{c}\mathbb{T}, \mathbf{d} = \mathbf{g}\mathbf{D} \tag{6.15}$$

At first, from $\mathbb{T}\mathbf{A} - \mathbf{G}\mathbb{T} = \mathbf{L}\mathbf{C}$, $\mathbf{L} = -\begin{bmatrix} \alpha_{s,0} \\ \vdots \\ \alpha_{s,s-1} \end{bmatrix}$ is determined. Let $\mathbf{c} = [0, \cdots, 1] \in \mathbb{R}^{1\times s}$. Taking $\mathbf{g}\mathbf{C} = \mathbf{c}\mathbb{T}$ gives $\mathbf{g} = \alpha_{s,s} \in \mathbb{R}^{1\times m}$. Using

$$\begin{bmatrix} \mathbf{H} \\ \mathbf{d} \end{bmatrix} = \begin{bmatrix} \mathbf{TB} - \mathbf{LD} \\ \mathbf{gD} \end{bmatrix} = \begin{bmatrix} \begin{bmatrix} \alpha_{s,1} & \alpha_{s,2} & \cdots & \alpha_{s,s} \\ \alpha_{s,2} & \alpha_{s,3} & \cdots & 0 \\ \vdots & \vdots & \vdots & \vdots \\ \alpha_{s,s} & 0 & \cdots & 0 \end{bmatrix} \begin{bmatrix} \mathbf{CB} \\ \mathbf{CAB} \\ \vdots \\ \mathbf{CA}^{s-1}\mathbf{B} \end{bmatrix} + \begin{bmatrix} \alpha_{s,0} \\ \vdots \\ \alpha_{s,s-1} \end{bmatrix} \mathbf{D} \\ \alpha_{s,s}\mathbf{D} \end{bmatrix}$$

$$= \begin{bmatrix} \alpha_{s,0} & \alpha_{s,1} & \cdots & \alpha_{s,s-1} & \alpha_{s,s} \\ \alpha_{s,1} & \alpha_{s,2} & \cdots & \alpha_{s,s} & 0 \\ \vdots & \cdots & \cdots & \vdots & \vdots \\ \alpha_{s,s} & 0 & \cdots & \cdots & 0 \end{bmatrix} \begin{bmatrix} \mathbf{D} \\ \mathbf{CB} \\ \mathbf{CAB} \\ \vdots \\ \mathbf{CA}^{s-1}\mathbf{B} \end{bmatrix} = \begin{bmatrix} \alpha_s \mathbf{H}_{u,s}(:, 1:l) \\ \alpha_s \mathbf{H}_{u,s}(:, l+1:2l) \\ \vdots \\ \alpha_s \mathbf{H}_{u,s}(:, sl+1:(s+1)l) \end{bmatrix}$$

then, $\mathbf{H} = \begin{bmatrix} \boldsymbol{\beta}_{s,0} \\ \vdots \\ \boldsymbol{\beta}_{s,s-1} \end{bmatrix} \in \mathbb{R}^{s\times l}$ and $d = \boldsymbol{\beta}_{s,s}$ are obtained. Different from the parity space-based residual generator in Eq. (6.10), the DO-based one can be used only requiring the current measurements. Furthermore, it has been proven that the two kinds of methods are equivalent when using the same parity vector. Therefore, for DO-based method, $r(k) \sim \mathcal{N}(0, \Sigma_r)$ with $\Sigma_r = \frac{\boldsymbol{\alpha}_{s,s}\mathbf{R}_{33}\mathbf{R}_{33}^T\boldsymbol{\alpha}_{s,s}^T}{N-1}$. T^2 test statistic in Eq. (6.11) is reduced to

$$T_{\mathbf{r}}^2 = \mathbf{r}^T \Sigma_{\mathbf{r}}^{-1} \mathbf{r} \sim \chi_1^2 \tag{6.16}$$

The threshold is given by $J_{th,T_r^2} = \chi_{1,\alpha}^2$. Then the decision logic is

$$\begin{cases} T_{\mathbf{r}}^2 \leq J_{th,T_{\mathbf{r}}^2} \Rightarrow \text{fault} - \text{free} \\ T_{\mathbf{r}}^2 > J_{th,T_{\mathbf{r}}^2} \Rightarrow \text{A fault occurs} \end{cases} \tag{6.17}$$

6.2 KPI-based FD using DO-based method

To address the KPI-based FD issue, the similar DO model can be established between the KPI data $\boldsymbol{\Theta}$ and process data $\mathbf{Y}$ as

$$\begin{aligned} \mathbf{x}_d(k+1) &= \mathbf{G}\mathbf{x}_d(k) + \mathbf{H}\mathbf{y}(k) + \mathbf{L}\theta(k) \\ \mathbf{r}(k) &= \mathbf{g}\boldsymbol{\theta}(k) - \mathbf{c}\mathbf{x}_d(k) - \mathbf{d}\mathbf{y}(k) \end{aligned} \tag{6.18}$$

Since the KPI measurements are online unavailable, the DO-based method cannot be directly designed. Assuming that $\boldsymbol{\theta} \in \mathbb{R}$, then g will be a scalar. Let $g = 1$, in this sense, the DO model (6.18) can be converted to a prediction model for KPI as

$$\begin{aligned} \mathbf{x}_d(k+1) &= (\mathbf{G}+\mathbf{Lc})\,\mathbf{x}_d(k) + (\mathbf{H}+\mathbf{Ld})\,\mathbf{y}(k) \\ \hat{\boldsymbol{\theta}}(k) &= \mathbf{c}\mathbf{x}_d(k) + \mathbf{d}\mathbf{y}(k) \end{aligned} \tag{6.19}$$

It is further simplified as

$$\begin{aligned} \mathbf{x}_d(k+1) &= \bar{\mathbf{G}}\mathbf{x}_d(k) + \bar{\mathbf{H}}\mathbf{y}(k) \\ \hat{\boldsymbol{\theta}}(k) &= \bar{\mathbf{c}}\mathbf{x}_d(k) + \bar{\mathbf{d}}\mathbf{y}(k) \end{aligned} \tag{6.20}$$

On the assumption that the KPI measurements are in steady state, then $\mathrm{E}\left(\hat{\boldsymbol{\theta}} - \bar{\boldsymbol{\theta}}\right) = 0$ holds, where $\bar{\boldsymbol{\theta}}$, the mean of $\boldsymbol{\theta}(k)$, is known and identified in terms of the recorded data. For this case, the T^2 test statistic is designed as

$$T_\theta^2 = \left(\hat{\boldsymbol{\theta}}(k) - \bar{\boldsymbol{\theta}}\right)^2 \Big/ \hat{\sigma} \tag{6.21}$$

where $\hat{\sigma}$ denotes the variance of $\hat{\boldsymbol{\theta}} - \bar{\boldsymbol{\theta}}$ and is offline estimated. The threshold is the same with Eq. (6.16). Finally, the decision logic is

$$\begin{cases} T_\theta^2 \leq J_{th,\theta} \Rightarrow \text{KPI is fault} - \text{free} \\ T_\theta^2 > J_{th,\theta} \Rightarrow \text{KPI is faulty} \end{cases} \tag{6.22}$$

with $J_{th,\theta} = \chi^2_{1,\alpha}$.

Remark 6.1. *For the case with multiple KPI, the DO-based method can be simply extended by establishing the DO model between each KPI and process measurements.*

6.3 KPI-based FD using subprocess-based method

Considering the subprocess embedded in the plant model, the relationship between KPIs and all process variables can be expressed as [12]

$$\boldsymbol{\theta} = \begin{bmatrix} \boldsymbol{\theta}_1 \\ \vdots \\ \boldsymbol{\theta}_l \end{bmatrix} = \boldsymbol{\Psi}_d \begin{bmatrix} \begin{bmatrix} \mathbf{u}_{s_1}^{(1)} \\ \mathbf{y}_{s_1}^{(1)} \end{bmatrix} \\ \vdots \\ \begin{bmatrix} \mathbf{u}_{s_M}^{(M)} \\ \mathbf{y}_{s_M}^{(M)} \end{bmatrix} \end{bmatrix} + \boldsymbol{\zeta} = \boldsymbol{\Psi}_d \mathbf{z}_s + \boldsymbol{\zeta} \tag{6.23}$$

where $s_1 ... s_M$ denotes time delay for each subprocess. Assume that in subprocess i, $s_i > n_i$ with n_i representing the order of the subprocess. For this problem, it is clear that a direct least-squares solution is not sufficient, since $\mathbf{u}$ and $\mathbf{y}$ are coupled with each other in different subprocess and such a solution will not accurately capture the dynamics of the whole system. Motivated by the parity space methods in the model-based and data-driven-based methods [2, 98], it is useful to define an instrument matrix that can successfully reflect the dynamic information of each subsystem.

To the end, the M residuals are collected: $\boldsymbol{\Upsilon} = \begin{bmatrix} \mathbf{r}^{(1)} \\ \vdots \\ \mathbf{r}^{(M)} \end{bmatrix}$, which is then used as the instrument variables. Due to the complexity in the whole system, $\mathbf{r}^{(i)}$s are mutually correlated.

Let the transformation matrix $\mathcal{I}$ be defined as $\begin{bmatrix} \mathcal{K}_{d,s_1}^{(1)} & & 0 \\ & \ddots & \\ 0 & & \mathcal{K}_{d,s_M}^{(M)} \end{bmatrix}$.

This gives

$$\hat{\boldsymbol{\Theta}}_d = \hat{\boldsymbol{\Psi}}_d \boldsymbol{\mathcal{Z}}_s = \hat{\boldsymbol{\Psi}}_{d,\Upsilon} \boldsymbol{\Upsilon} = \hat{\boldsymbol{\Psi}}_{d,\Upsilon} \mathcal{I} \boldsymbol{\mathcal{Z}}_s \tag{6.24}$$

Thus, the design of this method can be divided into two steps: (1) calculating the instrument variable for each subsystem and (2) monitoring the whole process based on the static model between the instrument variables and the KPI. In step (2), the static model is established using the LS-based method introduced in Chapter 4, where $\hat{\mathbf{\Psi}}_{d,\Upsilon} = [\mathbf{R}_\Upsilon \; 0] \begin{bmatrix} \mathbf{Q}_{\Upsilon,1} \\ \mathbf{Q}_{\Upsilon,2} \end{bmatrix}$. $\mathcal{P}_\Upsilon = \mathbf{Q}_{\Upsilon,1} {\mathbf{Q}_{\Upsilon,1}}^T$. Let $\mathbf{\Upsilon}_\theta = \mathcal{P}_\Upsilon \mathbf{\Upsilon}$, the T^2 test statistic is:

$$T^2_{d,\theta} = \mathbf{\Upsilon}_\theta^T \Sigma^{-1}_{\mathbf{\Upsilon}_\theta} \mathbf{\Upsilon}_\theta \tag{6.25}$$

In the dynamic KPI case, the decision is made based on

$$\begin{cases} T^2_{d,\theta} \leq J_{th,d} \Rightarrow \text{KPI is fault} - \text{free} \\ T^2_{d,\theta} > J_{th,d} \Rightarrow \text{KPI is faulty} \end{cases} \tag{6.26}$$

with $J_{th,d} = \chi^2_{l,\alpha}$.

Remark 6.2. *The subprocess-based method requires a large effort when creating the kernel realisations, system order n_i, and delay factor s_i for each of the M subsystems.*

6.4 Simulation results

For the purpose of understanding the behavior of the DO-based FD methods, the following system will be considered:

$$\begin{aligned} &\text{Process model} : \mathbf{y}(s) = G_u(s)\,\mathbf{u}(s) + G_\nu(s)\,\boldsymbol{\nu}(s) \\ &\text{KPI model} : \begin{cases} \mathbf{x}(k+1) = \mathbf{A}\mathbf{x}(k) + \mathbf{B}\begin{bmatrix} \mathbf{u}(k) \\ \mathbf{y}(k) \end{bmatrix} + \boldsymbol{\eta}(k) \\ \boldsymbol{\theta}(k) = \mathbf{C}\mathbf{x}(k) + \boldsymbol{\zeta}(k) \end{cases} \end{aligned} \tag{6.27}$$

where $G_u(s) = \frac{1.57}{61s+1}$, $G_\nu(s) = \frac{1}{50s+1}$. $\mathbf{A} = \begin{bmatrix} 0 & 1 & 0 \\ 0 & 0 & 1 \\ 0 & -0.02 & -0.5 \end{bmatrix}$, $\mathbf{B} = \begin{bmatrix} 1 & 0.5 \\ 0 & 1 \\ 1 & 1 \end{bmatrix}$, $\mathbf{C} = \begin{bmatrix} 1 & 1 & 1 \end{bmatrix}$, $\mathbf{x}(0) = \begin{bmatrix} 0 \\ 0 \\ 0 \end{bmatrix}$, $\boldsymbol{\eta}(k) \sim \mathcal{N}_3(0, \text{diag}(0.1, 0.1, 0.1))$, $\boldsymbol{\zeta}(k) \sim \mathcal{N}(0, 0.1)$, $\boldsymbol{\nu}(t) \sim \mathcal{N}(0, 0.2)$.

1000 samples are generated to give $\mathbf{U} \in \mathbb{R}^{1\times 1000}$, $\mathbf{Y} \in \mathbb{R}^{1\times 1000}$, and $\mathbf{\Theta} \in \mathbb{R}^{1\times 1000}$. Using these results, the KPI-based FD using DO-based method can be built, and parameters of interest are specified as $s = 10$, $n = 3$.

Three different scenarios will be considered:

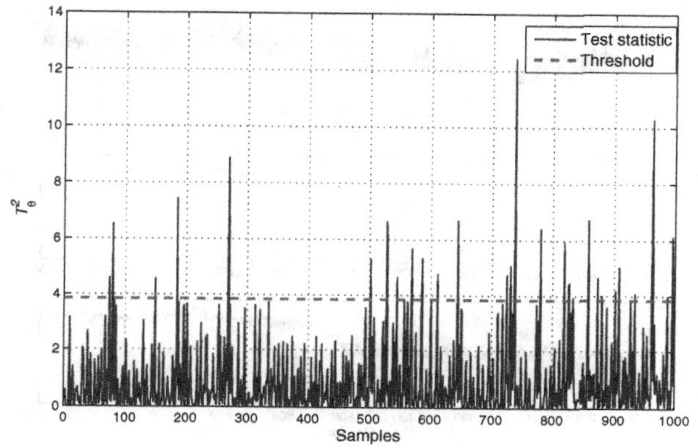

Figure 6.1: Detection performance for Scenario 1

- Scenario 1:
  Fault-free case with $\mathbf{u}(k) = \mathcal{N}(0,1)$.
- Scenario 2:
  Actuator fault with $\mathbf{u}_f(k) = \begin{cases} \mathcal{N}(0,1) & k \leq 500 \\ \mathcal{N}(5,1) & 500 < k \leq 1000 \end{cases}$
- Scenario 3:
  Sensor fault with $\mathbf{y}_f(k) = \begin{cases} \mathbf{y}(k) & k \leq 500 \\ \mathbf{y}(k) + 2 & 500 < k \leq 1000 \end{cases}$

Figure 6.1 shows the detection results for Scenario 1. We can see that the method could successfully track the behavior of KPI in fault-free case. In Scenario 2, actuator fault is simulated. From Figure 6.2, it can be observed that the KPI has been affected. Figure 6.3(a) gives the detection results for Scenario 2. It is shown that the detection is achieved quickly as the occurrence of the fault. By contrast, Figure 6.3(b) shows the results of DO-based method considering that the KPI is online available. It can seen that both methods provide similar performance.

In Scenario 3, a sensor fault is simulated to occur at 501rd sample. As shown in Figure 6.4, the process output $\mathbf{y}$ and the KPI $\boldsymbol{\theta}$ are affected. Figure 6.5(a) shows the results, where the DO-based method can instantly detect this fault. When the online KPI was used, the DO-based

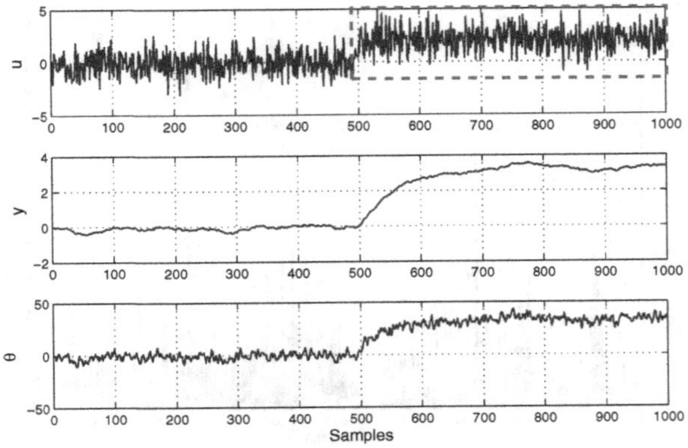

Figure 6.2: Profile of variables in Scenario 2

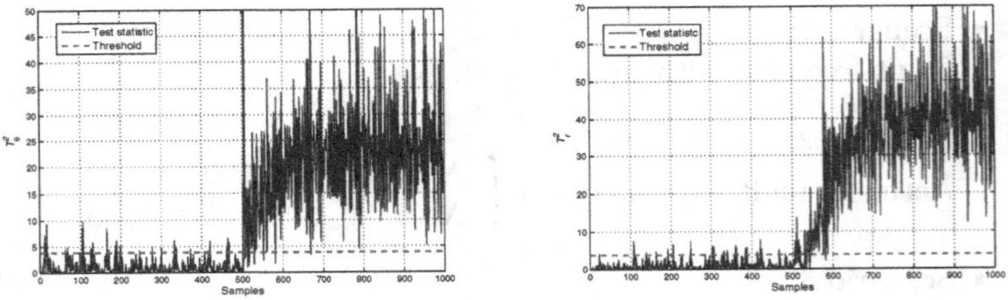

(a) DO-based FD results

(b) DO-based FD results when KPI is available

Figure 6.3: Fault detection performance for fault Scenario 2

method performs more effectively as shown in Figure 6.5(b). Therefore, we conclude that DO-based method can effectively identify the faults affecting KPI for this dynamic process.

6.5 Conclusions

In this chapter, KPI-based PM-FD methods that using the state-space to resolve the process dynamics have been discussed. Two methods were taken into account. The first one is developed based on the data-driven

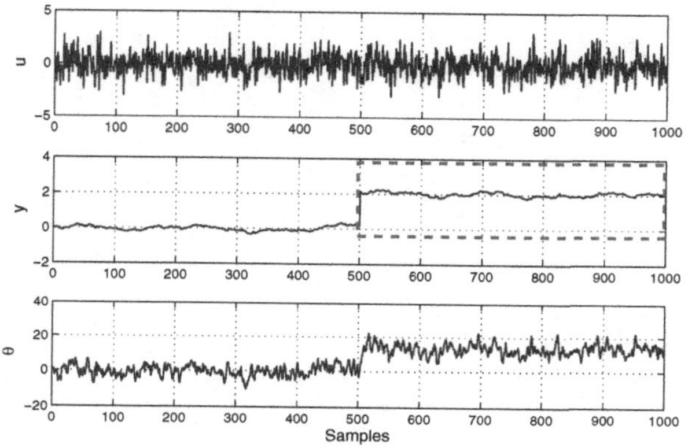

Figure 6.4: Profile of variables in Scenario 3

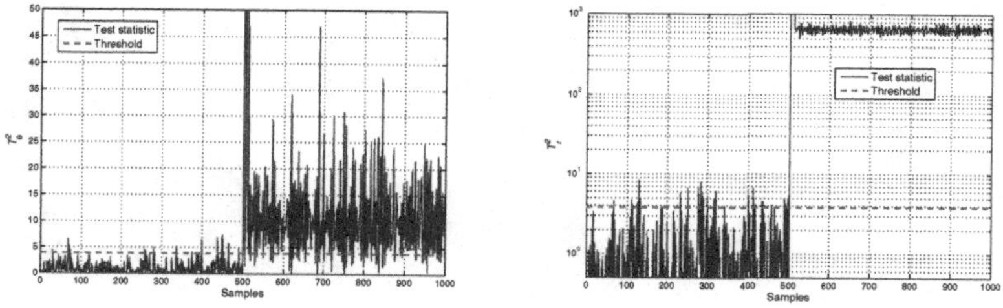

(a) DO-based FD results

(b) DO-based FD results when KPI is available

Figure 6.5: Fault detection performance for fault Scenario 3

realization of the diagnostic observer. By converting the DO to a soft-senor for predicting KPI, the status of KPI can be tracked using the residual between the predicted KPI and the mean value of it that is offline acquired. The other method considers the problem where there are a lot of subprocesses interconnected with each other and the KPI is dynamically related to process inputs and outputs. The method transfers the dynamic relationship to the static relationship between the KPI and the residual of each subprocess, and finally applies static method to detect faults.

7 Benchmark study and industrial application

In order to demonstrate and illustrate the practical applicability of the PM-FD methods studied in last chapters, in this chapter, the TE process and an industrial HSMR process are used to demonstrate the results achieved in previous chapters. As the TE process is a simulation for a chemical plant that longtime works in the stable state, it will be adopted to examine DPLS-based methods discussed in Chapter 5. As the HSMR process is a strongly dynamic process, the methods introduced in Chapter 6 are applied to this process.

7.1 Case studies on TE process

In this section, the TE process serves as the industrial benchmark to assess the capability of the DPLS-based methods for KPI-based PM-FD.

7.1.1 A brief introduction to TE process

The detailed piping and instrumentation diagram of TE process is shown in Figure 7.1 [49, 110], where there are five major operation units: a reactor, condenser, compressor, separator, and stripper. There are five gaseous inputs, namely A, B, C, D and E, fed into the reactor, where two liquid products G and H as well as a by-product F are formed. The products from the reactor firstly go through the condenser to cool the steam, then arrive at the separator. The condensed part moves to the stripper where the remaining reactants are removed, while the uncondensed part is compressed back to the reactor. Finally, the by-products are purged out of the system after the separator.

The process covers 22 process variables and 12 manipulated variables as briefly summarised in Table 7.1, where the variable tags keep the original definition in [49]. The process variables reflect the process running

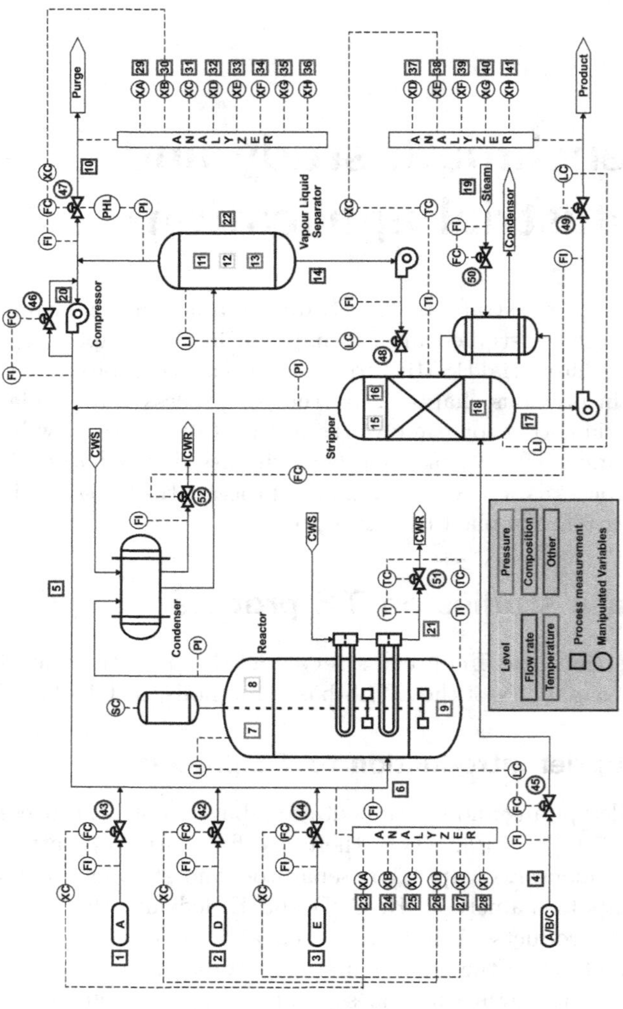

Figure 7.1: Schematic description of the TE process

environment and the content of components A-H in different units, while the manipulated variables control the amount of input reactants, the velocity of flow, and the agitator speed *etc.* For a more detailed description,

Table 7.1: Process and manipulated variables of TE process

Block	Description	Variable tag	
		Process	Manipulated
Feeds & Reactor	A feed (stream 1)	XMEAS(1)	XMV(3)
	D feed (stream 2)	XMEAS(2)	XMV(1)
	E feed (stream 3)	XMEAS(3)	XMV(2)
	A and C feed (stream 4)	XMEAS(4)	XMV(4)
	Compressor work	XMEAS(20)	—
	Compressor recycle valve	—	XMV(5)
	Recycle flow (stream 8)	XMEAS(5)	—
	Reactor feed rate (stream 6)	XMEAS(6)	—
	Reactor pressure	XMEAS(7)	—
	Reactor level	XMEAS(8)	—
	Reactor temperature	XMEAS(9)	—
	Reactor cooling water outlet temperature	XMEAS(21)	—
	Reactor cooling water flow	—	XMV(10)
	Agitator speed	—	XMV(12)
Condenser & Separator	Separator temperature	XMEAS(11)	—
	Separator level	XMEAS(12)	—
	Separator pressure	XMEAS(13)	—
	Separator underflow (stream 10)	XMEAS(14)	XMV(7)
	Condenser cooling water outlet temperature	XMEAS(22)	—
	Condenser cooling water flow	—	XMV(11)
	Purge rate (stream 9)	XMEAS(10)	—
	Purge valve (stream 9)	—	XMV(6)
Stripper	Stripper level	XMEAS(15)	—
	Stripper pressure	XMEAS(16)	—
	Stripper underflow (stream 11)	XMEAS(17)	XMV(8)
	Stripper temperature	XMEAS(18)	—
	Stripper steam flow	XMEAS(19)	XMV(9)

the reader is referred to [49, 110] and the simulator is downloadable at the website[1]. This simulator contains the closed-loop control scheme and allows an easy setting of the operation modes, measurement noises, sampling time. This makes it convenient to support the Monte-Carlo experiment for evaluating the considered approaches in the statistical framework. The operation conditions of the study are: the sampling time is 3 minutes and the process runs continuously for 100 h (2000 samples) in both fault-free and faulty episodes. The component variables are observed with a time delay varying from 6 to 15 min. As the time delay is not the main focus, thus, will not be stressed in this thesis. Relevant information can be found in [8]. The recorded data $\mathbf{Y}_{obs}$ include 9 manipulated variables and 22 process variables, *i.e.*

$$\mathbf{y}_{obs} = [\text{xmeas}\,(1-22)\,,\text{xmv}\,(1-4)\,,\text{xmv}\,(6-7)\,,\text{xmv}\,(10-11)]^T \in \mathbb{R}^{31}$$

where xmv(5), xmv(9) and xmv(12) are excluded since they are invariant during the simulation.

[1] http://depts.washington.edu/control/LARRY/TE/download.html

To indicate the operation performance of this process, an objective function dedicated to the operating cost is defined as

$$\boldsymbol{\theta}_{obs} = D_p V_p + D_{pro} V_{pro} + D_{comp} W_{comp} + D_s V_s \tag{7.1}$$

where D_P, D_{comp} and D_s the costs for the operating of the purge, compressor and steam; D_{pro} represents the product stream cost; V_p, V_{pro} and V_s are corresponding operating rate; and W_{comp} is the work that the compressor requires [49]. We can see that the status of KPI for this process is primarily determined (1) by the loss of raw materials, *i.e.* the loss of the raw material in the purge gas, the product stream and by means of the two side reactions; (2) by the costs of the raw materials and the products leaving in the purge stream, the raw materials leaving in the product stream, and using an assigned cost to the amount of F formed; (3) the costs of the compressor work and steam to the stripping column. Downs *et al.* have given the calculable formula for each part of cost [49], where it can be found that the KPI is not available in time, since the component cost is generally delayed by 6 to 15 min. Thus, it is necessary to perform the KPI-based PM-FD for supervising the KPI's status in TE process. During the training phase, the significance level is set as $\alpha = 0.05$.

7.1.2 Results and discussion

We first consider the normal operating operation. The application of the two PM-FD methods leads to the results shown in Figure 7.2. It can be observed that both of them can provide acceptable FARs. This suggests that the two DPLS-based methods have been properly configured for this process and can work with the no fault case.

The benchmark has 20 predefined process faults [49], of which # 1 to 7 and 13 are additive faults. Based on the previous study [71], fault 3 cannot influence the KPI, thus, it is not included in this simulation. Note that fault 13 is assumed to a constant fault in this simulation. Before calculating EDD, the fault vector $\mathbf{f}$ is estimated using $\mathbf{f} \approx \frac{1}{N_f} \sum_{i=1}^{N_f} \mathbf{y}_{f,i}$. Using Eqs. (2.15), (5.22) and (5.23), EDD for all considered faults can be obtained and shown in Table 7.2. It can be seen that for faults 1, 2, 6, 13, the EDD given by both DPLS methods is zero, which means that the two methods can effectively detect these faults without time delay. The

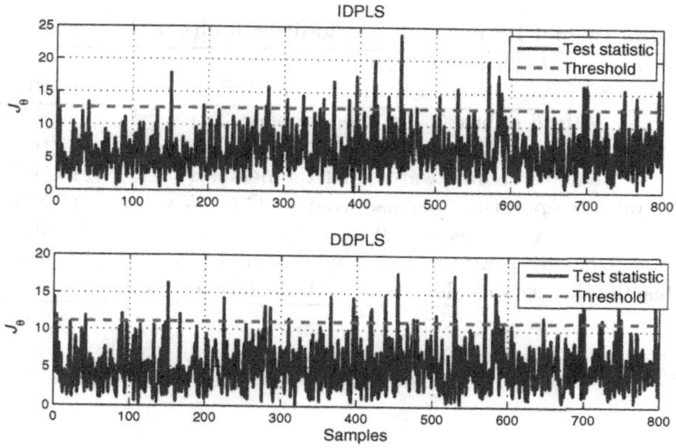

Figure 7.2: Detection performance for fault-free case using two DPLS methods

result is further verified by Figure 7.3, the detection performance for fault 1, where it can be found that they both behave well and can detect this fault as soon as it occurs. It is worthwhile noting that for this fault, if the FDR is calculated using the numerical approximation-based method [40], the IDPLS method will be thought to be better. Therefore, EDD can accurately assess FM-FD performance. The two methods give different EDD for fault 7. The IDPLS method performs better than the second one, and hence is suggested to detect this type of fault. Figure 7.4 shows the probability distribution and the cumulative density function of the detection delay for detecting fault 7. The probability value is calculated using the formulae presented in Eqs. (5.22) and (2.19). It can be seen that as the detection delay takes larger values, the associated probability value decreases quickly. When DD takes around 25 to 30, the CDF of DD approximate 1. In Figure 7.4, the difference between the two methods is that using IDPLS, the probability the DD takes small values is large than obtained using DDPLS. This could account for the different EDD provided by them. For faults 4 and 5, both methods show poor EDD performance, which implies that it will take them a long time period to detect these faults. Figure 7.5 shows the probability distribution of DD for fault 4. In this case, compared against Figure 7.4, it can be

Table 7.2: EDD of two DPLS methods for additive faults in TE process

Fault #	Fault information Description	EDD IDPLS	 DDPLS
1	A feed reduce 3%, C feed increase 3%	0	0
2	B feed increase 5%	0	0
4	Reactor cooling water inlet temperature	16.5076	16.4438
5	Inlet temperature increase from 40°C to 45°C.	17.3597	17.8501
6	A feed reduces 0 from 0.2505	0	0
7	C header pressure loss-reduced availability	0.1085	0.9465
13	Reaction kinetics:random walk with deviation of 0.25.	0	0

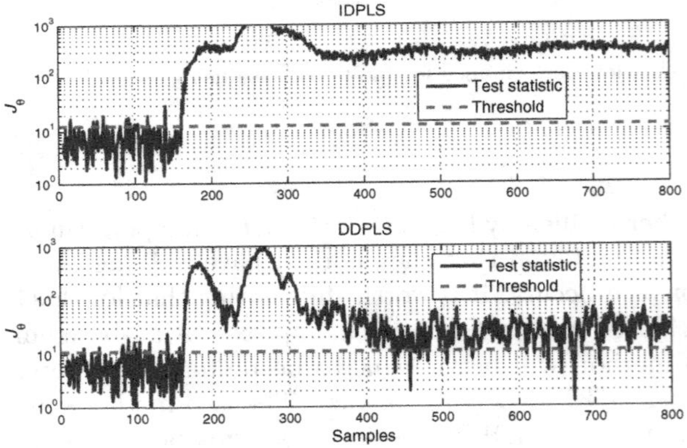

Figure 7.3: Detection of fault 1 using two DPLS methods

found that the PDF value decreases slowly, and the CDF value likewise increases slowly.

The behavior of EDD in this simulation is similar to that shown in the numerical example. The PM-FD performance of two DPLS methods strongly depends on the applied process and how a fault impact KPIs. Thus, a better selection of the most suitable method should reply on as much as possible on prior process and fault information.

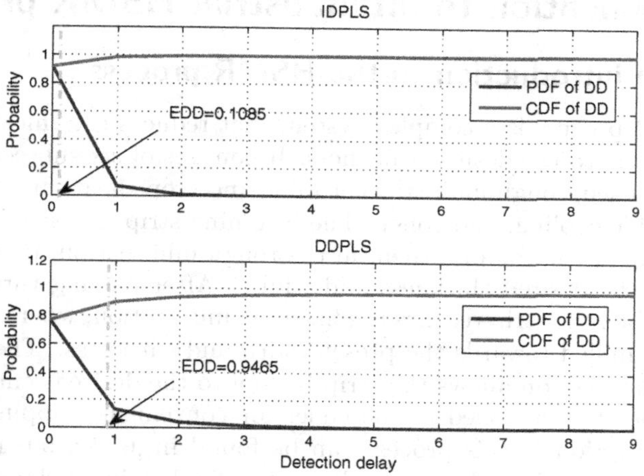

Figure 7.4: Probability distribution of DD for fault 7 using two DPLS methods

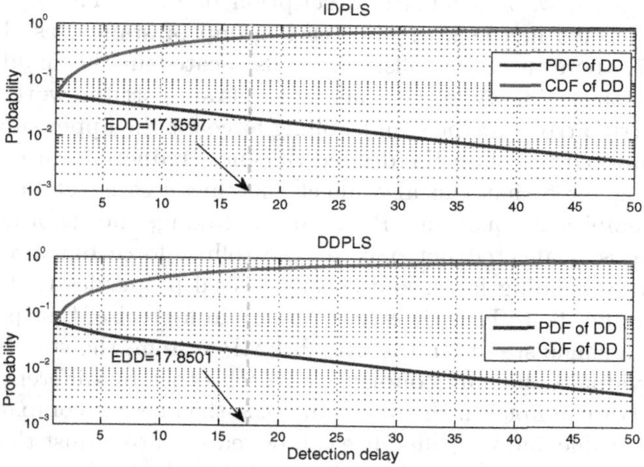

Figure 7.5: Probability distribution of DD for fault 4 using two DPLS methods.

7.2 Application to an industrial HSMR process

7.2.1 An introduction to the HSMR process

The HSMR process is a complex system that reduces the thickness of a hot strip steel to the desired thickness. It consists of six subsections: reheating furnace, rough mill, transfer table and crop shear, finishing mill, run-out table cooling, and coiler. The incoming strip is first reheated in the reheating furnace, and, then, in the rough mill section, it is roughly shaped to the desired thickness and width. After passing through the transfer table, the strip reaches the finishing mill section, where it will be accurately milled towards the preset width and thickness. The run-out table cooling section allows the strip to cool to the desired temperature. Finally, the strip is coiled in the coiler for convenient shipping. A detailed description of this process can be found in [6, 10, 99, 131]. For the HSMR, one of the key variables is the final strip thickness, which is primarily determined by the finishing mill process (FMP). Therefore, the focus of this example will be on analysing and understanding the PM-FD issue in FMP.

Figure 7.6 shows a schematic description of FMP. There are seven groups of stands in the FMP. For each group of stands, as shown in Figure 7.7, there are four rolls: two rolls located in the middle work directly on the strip, while the other two rolls support the working ones. Before a strip arrives at the stand, rolling force for the upper supporting roll is computed based on the desired thickness reduction rate. As well, the bending force that can also affect the thickness is set beforehand using an empirical equation. Physically speaking, the deformation of the thickness is affected not only by the rolling force but also by the temperature, bending force, and other physical properties that depend on the specific steel. Thus, it is hard to obtain a precise, first principles model for a single stand. In the overall FMP system, the stands do not work individually, but are coupled with each other by different control schemes. For example, in the last stand, the thickness is compared with the desired value and the difference can be fed back to adjust the milling force in that or previous stands. It is noted that the thickness cannot be measured between two stands, instead the gap measurements between two working rolls are available. Due to the rebounding phenomenon, the thickness is approximately equal to the gap less the impact of the roll's stiffness. However, since the stiffness is hard to precisely calculate, it

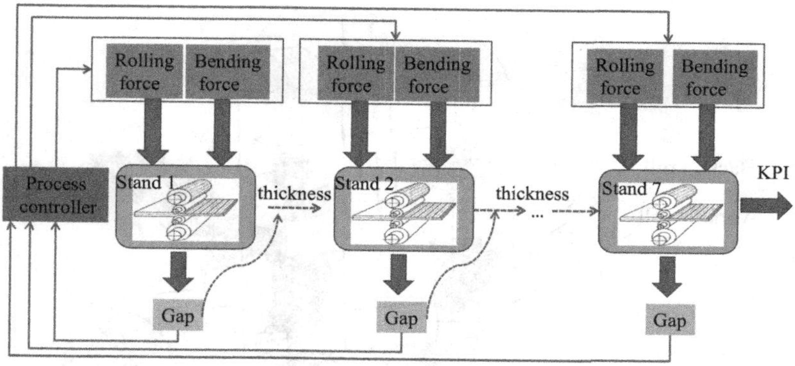

Figure 7.6: Schematic description of a large-scale FMP

is impossible to adjust the downstream stands based on the upstream thickness. In HSMP, the most important KPI is the thickness at the exit of the multistand system. It is measured using a X-ray device located at a distance from the stands, which will cause time delay in the feedback control system.

For each of the seven stands, the control structure is explicitly shown in Figure 7.6. The manipulated variables are the rolling and bending forces, while the gap is the controlled variable. The KPI, which has been defined as the final thickness, is closely linked with all the process variables. This suggests that implementing the dynamic PM-FD systems introduced in Chapter 6 would be appropriate. Two different strip thicknesses will be considered: E(KPI) = 2.70 mm and E(KPI) = 3.95 mm. The training data set consists of 3,000 data points collected to form the data structure $\{\mathbf{U}^{(i)} \in \mathbb{R}^{2\times3000}, \mathbf{Y}^{(i)} \in \mathbb{R}^{1\times3000}\}$ with $i = 1,..,7$, and $\mathbf{\Theta} \in \mathbb{R}^{1\times3000}$. Unlike the DO-based method, for the subprocess-based method, the kernel representation vector $\mathcal{K}_{d,s}$, needs to be developed for each control loop and the parameters s_i and n_i should be fixed beforehand. Since this step is not the main focus, the results of fixing these parameters are not shown. Additional information can be found in [2]. The results obtained are $s_i = 13$ with $i = 1,...,3$, $s_j = 14$ with $j = 4,...,7$, and $n_k = 5$ for $k = 1,...,7$.

Four different scenarios will be considered:

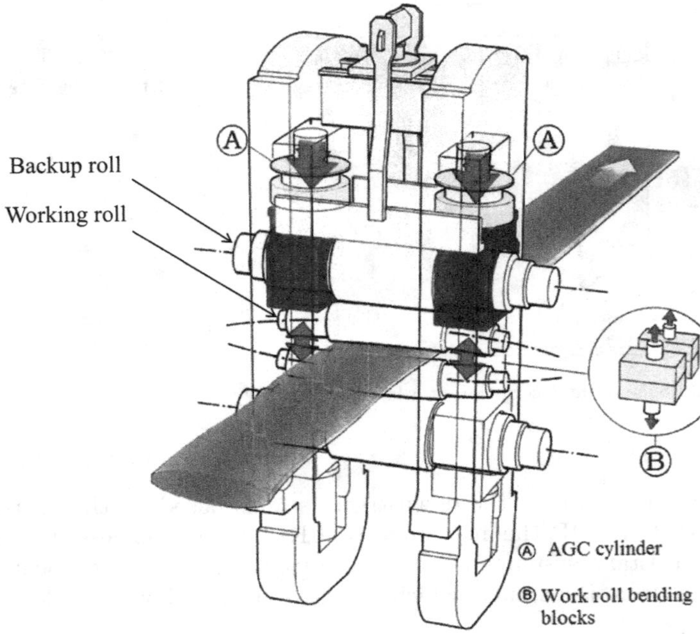

Figure 7.7: Schematic description of the stand in FMP

- Scenario 1: Fault-free case, that is, there are no faults in the system.
- Scenario 2: The control loop deteriorates in the 4th stand at 20 s.
- Scenario 3: The cooling water valve is blocked between the 2nd and 3rd stands at 12 s.
- Scenario 4: The bending force sensor in the 5th stand malfunctions at 10 s, but the KPI is not affected.

7.2.2 Results and discussion

In order to understand the behaviour of the residual signals for the subprocess-based method compared with the DO-based method, representative residual signals for both cases will be examined and compared

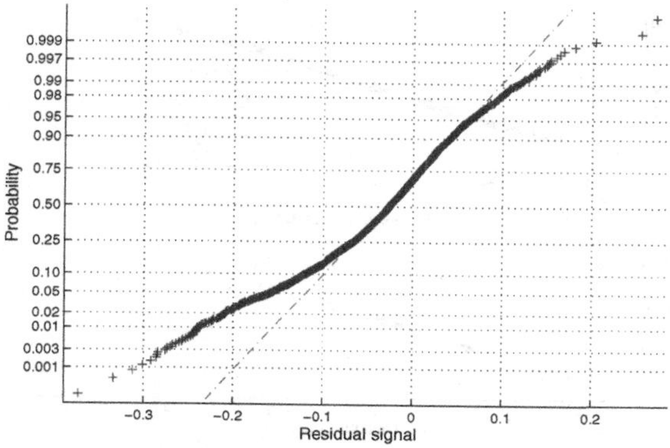

Figure 7.8: Normal distribution plot of residual signals using DO-based method

to how well they resemble the normal distribution. A normally distributed signal will lie along the straight line with minimal deviations from it. As shown in Figure 7.8, the residual signal from DO-based method has huge deviations at the left and right ends (tails) from the straight line. This deviation strongly suggests that the signal is not normal. Therefore, using this signal to monitor the process will lead to many false alarms, which is not desirable. On the other hand, Figure 7.9 shows two representative residual signals for two subprocesses. Here the data points all lie much closer to the straight line without huge deviations from it. Thus, this suggests that the results are normally distributed.

In Figure 7.10, we can see the results for Scenario 1. It can be observed that there are very few false alarms in $T^2_{d,\theta}$, even at the starting stage when the process is very unstable. For the strip with KPI = 3.9 mm, the result is similar with the FAR equalling 0.0410. Figure 7.11 gives the results for Scenario 2. It can be seen that the KPI has been affected, and detection is achieved before KPI is even affected. As described above, this fault occurred in the 4th stand. Thus, the 4th subprocess was instantly influenced and this was then transmitted to the downstream subprocesses until it finally reached the KPI. As shown in Figure 7.11, $T^2_{d,\theta}$ has performed very well in this case. By applying the popular relative contribution plot based method to the residuals of the 7 subprocesses [14, 59], mean contribution values of the first 50 samples

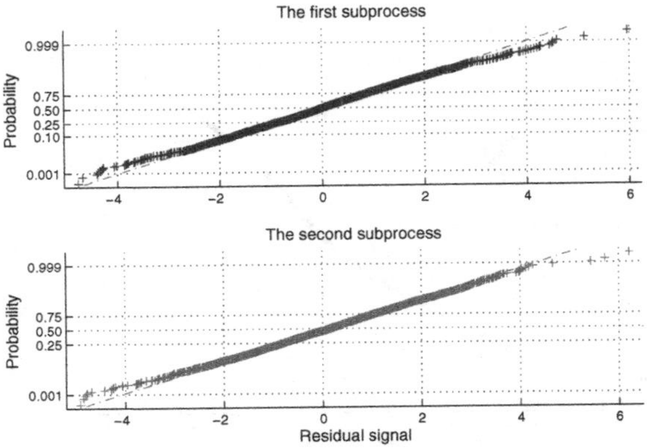

Figure 7.9: Normal distribution plot of residual signals for subprocess 1 and 2

are $\mathcal{C}_{s1} = 0.012, \mathcal{C}_{s2} = 0.008, \mathcal{C}_{s3} = 0.045, \mathcal{C}_{s4} = 0.822, \mathcal{C}_{s5} = 0.100, \mathcal{C}_{s6} = 0.003, \mathcal{C}_{s7} = 0.001$. It is evident that the 4th stand is the faulty origin, which is consistent with the definition of the fault. Figure 7.12 shows the performance of DO-based method for this scenario. In contrast with Figure 7.11, there more false alarms arisen during the onset and middle terms of the process, which are unacceptable in practice.

Figure 7.13 demonstrates the subprocess-based detection results for Scenario 3, which contains a cooling water fault with a strip whose KPI = 3.95 mm. This is also a KPI-relevant fault as can be seen in the change at 16 s. Similar to Figure 7.11, the fault was also detected before it had influenced the KPI. The fault occurred between the 2nd and 3rd stands and directly affect the 3rd subprocess at about 13 s. $T^2_{d,\theta}$ succeeded to detect it at 12.5 s, which shows an effective performance. After the fault has been manually eliminated at around 22.5^{th} s the detection showed some disturbance, which is caused by the switch of parameters in some controllers. Like in the last case, the relative contribution values are obtained: $\mathcal{C}_{s1} = 0.017, \mathcal{C}_{s2} = 0.032, \mathcal{C}_{s3} = 0.454, \mathcal{C}_{s4} = 0.306, \mathcal{C}_{s5} = 0.114, \mathcal{C}_{s6} = 0.073, \mathcal{C}_{s7} = 0.003$, which shows that the subprocess 3 is responsible for the fault. Figure 7.14 shows the DO-based results for this faut. Similar to Scenario 2, there exist undesired alarms.

Figure 7.15 shows the results for Scenario 4, which also considers a strip whose KPI = 3.95 mm. However, in this case, the KPI is not

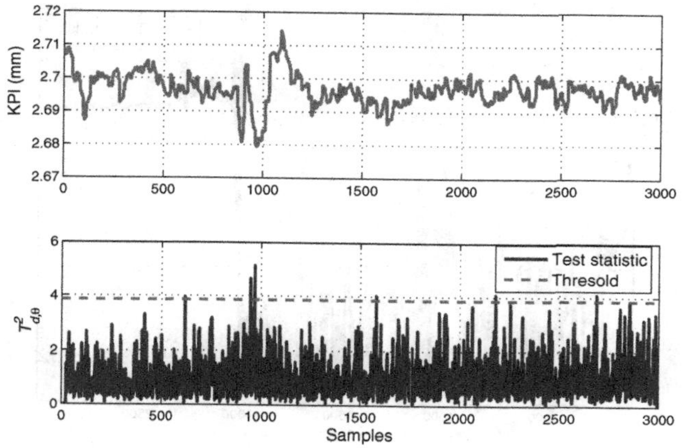

Figure 7.10: Monitoring result for Scenario 1

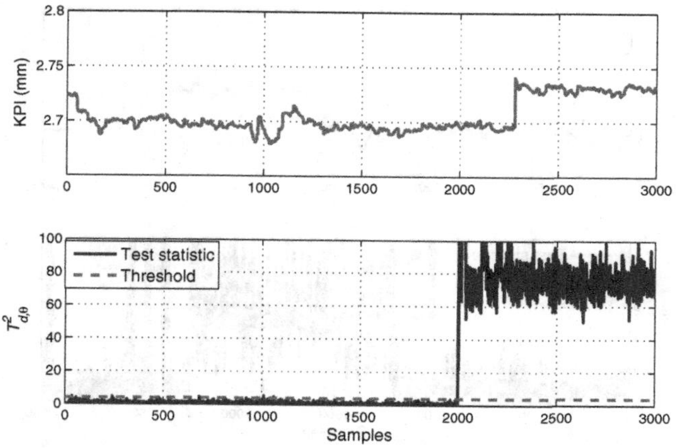

Figure 7.11: Monitoring result for Scenario 2

influenced by this fault. Figure 7.15 demonstrates consistent results with normal KPI measurements. $T^2_{d,\theta}$ returns a few false alarms (the rate is 0.054), which is acceptable compared to the significance level $\alpha = 0.05$.

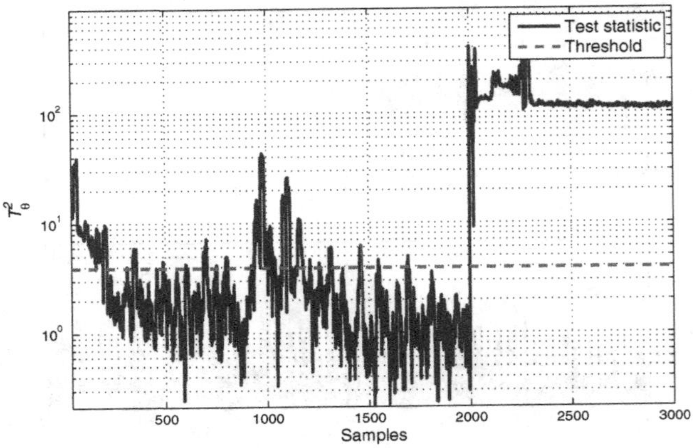

Figure 7.12: Monitoring result for Scenario 2 using DO-based method

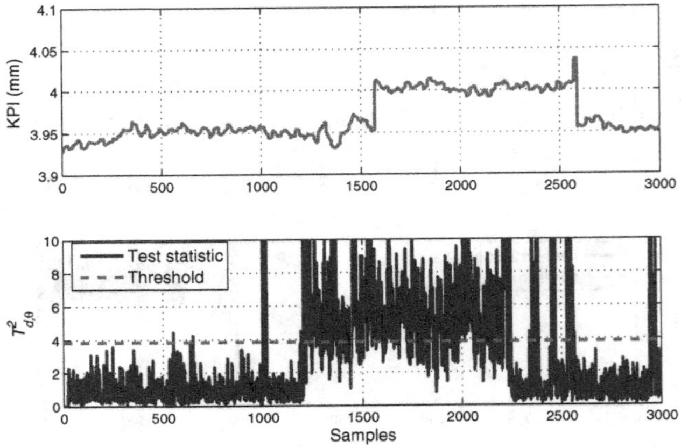

Figure 7.13: Monitoring result for Scenario 3

The results by DO-based methods show poor performance in Figure 7.16. As conclusion, we can see that the dynamic methods considered in Chapter 6 can effectively identify faults before they cause changes in the

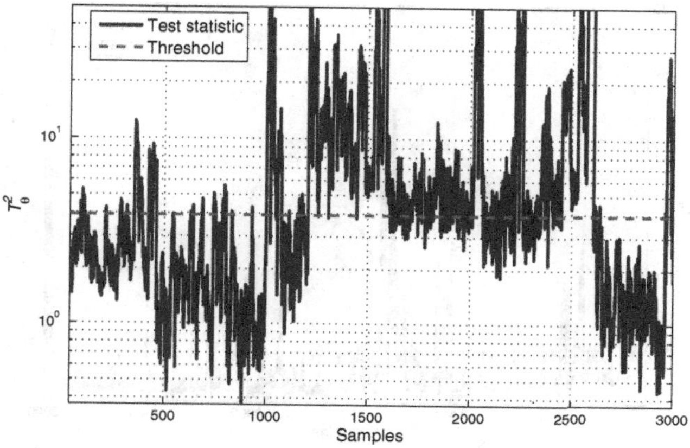

Figure 7.14: Monitoring result for Scenario 3 using DO-based method

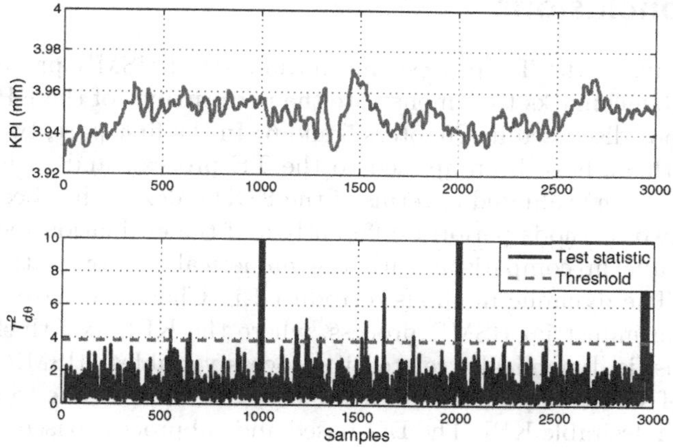

Figure 7.15: Monitoring result for Scenario 4

KPI of HSMR process. This allows for early and efficient handling of any potential problems. The subprocess-based method provides a better performance in FAR.

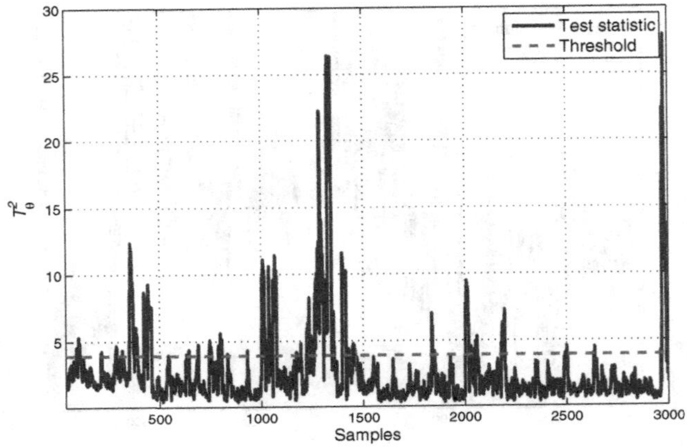

Figure 7.16: Monitoring result for Scenario 4 using DO-based method

7.3 Conclusions

In this chapter, the TE process and an industrial HSMR process have served as benchmarks to demonstrate the performance of the KPI-based FD methods discussed in Chapters 5 and 6. In the first part, two DPLS-based methods have been applied to the TE process, and their performance has been evaluated in terms of the EDD index. It has been found that the two methods perform differently and the evaluation results are more reliable in comparison with the numerical approximation-based method. The dynamic methods introduced in Chapter 6 have been applied to an industrial HSMR process, where the KPI, exit thickness of strip steels, is dynamically related to process variables. HSMR process consists of seven subprocesses that are connected in cascade to produce steels with desirable KPI. The DO-based and subprocess-based methods have been considered using real HSMP process data. It can be found that the subprocess-based method that takes the dynamics within each subprocess provided more efficient performance. By contrast, the DO-based one produces more false alarms which is not acceptable in industry.

8 Conclusions and future work

8.1 Conclusions

In many industrial systems, the collection of routine operating data becomes easy. This leads to a large amount of information-rich data that can be adopted for PM-FD. A typical use of them is, together with the online available process measurements, to track the behavior of KPI variable which are often online unavailable. Over recent decades, development of KPI-based PM-FD methods become to be an active area of research. On the other hand, there is currently lack of a framework for the comparison and assessment of existing methods. Such a framework has been presented in this thesis.

Firstly, the background material and motivations for the thesis have been introduced in Chapter 1. The growing industrial demands for plant KPIs require the development of KPI-based PM-FD methods such that the behavior of KPI can be timely tracked. From the theoretical viewpoint, there exist plenty of methods, while comparisons and interconnections among them have not attracted sufficient research attention. From an industrial perspective, performance of these methods deserve to be assessed under a unified framework to determine the selection of an appropriate method for a process.

In Chapter 2, mathematical descriptions of static and dynamic processes have been presented. Two types of faults, *i.e.* additive and multiplicative faults, have been considered including the mathematical representation and how they affect the process performance. Furthermore, The commonly used performance evaluation indices, FAR and FDR, have been dealt in detail. A new EDD index was proposed based on the statistical formula of FDR. Rigorous mathematical handling demonstrates that EDD can be applied to both additive and drift faults. Simulation study on a numerical case has verified that application of EDD yields more accurate and consistent results compared with the numerical approximation-based method.

Chapter 3 has focused on comparing the two widely used test statistics, J_{T^2} and J_Q, using the statistical as a tool. With the aid of χ^2 distribution, their nature to detect additive and multiplicative faults have been revealed. That is, detecting additive faults is equivalent to shifting the PDF of χ^2 to the right, which is achieved by increasing the noncentrality parameter, while detecting multiplicative faults is equivalent to shifting the threshold to the left. It has also been shown that in most cases, when detecting additive faults, J_{T^2} performs better than J_Q, in few cases that depend on properties of covariance matrices and the choice of $J_{th,Q}$, J_Q performs better. The drawback of them to detect multiplicative faults has been investigated, which are that both methods cannot reach FDR $= 1$. The drawback can be overcome by some advanced statistics which were reviewed in the last part of this chapter. Comparisons of them have been carried out with respect to calculation and FD performance. The overall theoretical comparison results for J_{T^2} and J_Q, as well as some intermediate results, has finally been tested by means of numerical simulations.

Based on the fundamental issues addressed in Chapters 1-3, Chapters 4-6 have been dedicated to the assessment the KPI-based PM-FD methods for static processes, dynamic processes that work in steady state, and dynamic processes in general. In Chapter 4, the commonly used KPI-based PM-FD methods for static processes have been thoroughly studied. The existing approaches have been, for the first time, sorted into three classes: the direct method, LS-based and PLS-based methods. Based on the theoretical analysis, their interconnections and computation costs have been investigated. Furthermore, EDD has been applied to evaluate their performance to detect offsetting, drift and multiplicative faults. A numerical simulation example has been used to demonstrate the results. It is worth once again noting that

Although PLS-based methods are most commonly encountered in KPI-based PM-FD area for static processes, it was not originally developed for this purpose. The recently improved methods (e.g., T-PLS and C-PLS) aim at solving and dealing with the specific issues in the model. However, the new formulation cause increase in the design efforts that calls for a need to perform further improvements. Indeed, rather than the post-modification of PLS, the pre-modification like LS-based methods seems cost-efficient and direct, which thus deserves more attention in this area.

When the considered process works in steady state, the DPLS models can be used to solve KPI-based PM-FD problems. Chapter 5 has concentrated on examining two popular DPLS methods: the DDPLS method, which computes the optimal decomposition based on individual direction vectors, and the IDPLS method, which computes the optimal decomposition based on the assumption that the direction vectors for all eigenvectors is the same. Comparisons between the their original formulas as well their NIPALS alternatives have been made. Their FD performance to KPI-based PM-FD was evaluated using EDD, where a numerical case was utilized to demonstrate the results.

Chapter 6 has presented the application of state space representation to describe the dynamic relations between process and KPI variables. A DO-based method has been introduced which are based on the idea of PS-based PM-FD. The other method addressed the problem where there are a lot of subprocesses interconnected with each other and the KPI is dynamically related to process inputs and outputs. The method transforms the dynamic relationship to the static relationship between the KPI and the residual of each subprocess, and finally applies static method to detect faults.

Finally, the developed theoretical results in Chapter 5 have been validated on a benchmark processes, where the results are consistent with the numerical simulation in Chapter 5. The methods addressed in Chapter 6 have been applied to a real industrial HSMR process to monitor the KPI, the final thickness of steel products.

8.2 Future work

Based on the achieved results, future work can be taken in multiple directions, such as:

- As shown in Chapter 3, T^2 and Q- statistics suffer inherent drawbacks for handling multiplicative faults. Thus, much work will be concerned to develop more efficient test statistics such that detection of this type of fault can be as easy as that for additive faults.

- Developing a performance assessment software based on the ED-D index and methods investigated in Chapters 4 and 5 will be accomplished in future.

- The KPI-based PM-FD methods considered in this thesis are limited to linear processes. Indeed, most industrial plants are full of nonlinearities and work in different operating modes [30, 31, 77]. For these cases, developing nonlinear PM-FD methods can gain more practical benefits.
- Besides the methods addressed in this work, there are a plethora of KPI-based PM-FD methods from machine learning, pattern recognition, imagine processing fields, *etc.* [132, 133]. Comparing them in both theoretical and practical performance aspects will be performed in the near future.
- Future work will also focus on addressing the fault isolation issue under the statistical framework [58–60, 134].

Bibliography

[1] S. X. Ding, *Model-Based Fault Diagnosis Techniques-Design Schemes Algorithms and Tools.* 1st ed. Berlin, Germany: Springer-Verlag, 2013.

[2] S. X. Ding, *Data Driven Design of Fault Diagnosis and Fault Tolerant Control Systems.* London, Great Britain: Springer, 2014.

[3] R. Isermann, *Fault Diagnosis Systems: An Introduction from Fault Detection to Fault Tolerance.* Berlin, Germany: Springer-Verlag, 2006.

[4] Z. Gao, C. Cecati, and S. X. Ding, "A survey of fault diagnosis and fault-tolerant techniqures part I: fault diangosis with model-based and signal-based approaches," *IEEE Trans. Ind. Electron.*, vol. 62, no. 6, pp. 3757–3767, 2015.

[5] Z. Gao, C. Cecati, and S. X. Ding, "A survey of fault diagnosis and fault-tolerant techniqures part II: Fault Diagnosis with Knowledge-Based and Hybrid/Active Approaches," *IEEE Trans. Ind. Electron.*, vol. 62, no. 6, pp. 3768–3774 , 2015.

[6] S. X. Ding, S. Yin, K. X. Peng, H. Y. Hao, and B. Shen, "A novel scheme for key performance indicator prediction and diagnosis with application to an industrial hot strip mill," *IEEE Trans. Ind. Inform.*, vol. 9, no. 3, pp. 39–47, 2013.

[7] H. Y. Hao, K. Zhang, S. X. Ding, A. Haghani, and H. Luo, "A data-based performance management framework for large-scale industrial processes, " in: *2014 IEEE Conf. on Control Applications*, Antibes, France, Oct. 2014, pp. 772–777.

[8] Y. A. W. Shardt, H. Y. Hao, and S. X. Ding, "A new soft-sensor-based process monitoring scheme incorporating infrequent KPI measurements," *IEEE Trans. Ind. Electron.*, vol. 62, no. 6, pp. 3843–3851, 2015.

[9] H. Y. Hao, K. Zhang, S. X. Ding, Z. W. Chen and Y. G. Lei, "A data-driven multiplicative fault diagnosis approach for automation processes," *ISA Trans.*, vol. 53, no. 3, pp. 1436–1445, 2014.

[10] K. X. Peng, K. Zhang, G. Li, and D. H. Zhou, "Contribution rate plot for nonlinear quality-related fault diagnosis with application to the hot strip mill process," *Control Eng. Pract.*, vol. 21, no. 4, pp. 360-369, 2013.

[11] Z. K. Hu, Z. W. Chen, W. H. Gui, and B. Jiang, "Adaptive PCA based fault diagnosis scheme in imperial smelting process," *ISA Trans.*, vol. 53, no. 5, pp. 1446–1455, 2014.

[12] G. Li, S. J. Qin, B. S. Liu, and D. H. Zhou, "Quality relevant data-driven modeling and monitoring of multivariate dynamic processes: the dynamic T-PLS approach," *IEEE Trans. Neural Netw.*, vol. 22, no. 12, pp. 2262–2271, 2011.

[13] S. J. Qin, "Survey on data-driven industrial process monitoring and diagnosis," *Annu. Rev. Control*, vol. 36, no. 2, pp. 220–234, 2012.

[14] S. J. Qin, "Statistical process monitoring: basics and beyond," *J. Chemometr.*, vol. 17, no. 8-9, pp. 480–502, 2003.

[15] T. Kourti, "Preface to John MacGregor Festschrift," *Ind. Eng. Chem. Res.*, Special Issue: John MacGregor Festschrift.

[16] Special Issue Honouring J. F. MacGregor on his 65th Birthday, *Can. J. Chem. Eng.* vol. 86, no. 5, pp. 813–970, 2008.

[17] Special Issue Honouring J. F. MacGregor on his 60th Birthday, *J. Chemometr.*, vol. 17, no. 1, pp. 1–109, 2009.

[18] J. MacGregor and A. Cinar, "Monitoring, fault diagnosis, fault-tolerant control and optimization: Data driven methods," *Comput. Chem. Eng.*, vol. 47, pp. 111–120, 2012.

[19] J. V. Kresta, J. F. MacGregor, and T. E. Marlin, "Multivariate statistical monitoring of process operating performance," *Can. J. Chem. Eng.*, vol. 69, pp. 35–47, 1991.

[20] J. F. Macgregor, C. Jaeckle, C. Kiparissides, and M.Koutoudi, "Process monitoring and diagnosis by multiblock PLS methods," *AIChE J.* vol. 40, no. (5), pp. 826–838, 1994.

[21] T. Kourti and J. F. MacGregor, "Process analysis, monitoring and diagnosis, using multivariate projection methods," *Chemometr. Intell. Lab. Syst.* vol. 28, no. 1, pp. 3–21, 1995.

[22] L. H. Chiang, E. L. Russell, and R.D. Braatz, *Fault Detection and Diagnosis in Industrial Systems in: Advanced Textbooks in Control and Signal Processing.* London, Great Britain: Springer-Verlag, 2001.

[23] G. Li, S. J. Qin, and D. H. Zhou, "Output relevant fault reconstruction and fault subspace extraction in total projection to latent structures models," *Ind. Eng. Chem. Res.*, vol. 49, no. 19, pp. 9175–9183, 2010.

[24] S. J. Qin and Y. Y. Zheng, "Quality-relevant and process-relevant fault monitoring with concurrent projection to latent structures," *AIChE J.*, vol. 59, no. 2, pp. 496–504, 2013.

[25] J. Mori and J. Yu, "Quality relevant nonlinear batch process performance monitoring using a kernel based multiway non-Gaussian latent subspace projection approach," *J. Process Control*, vol. 24, no. 1, pp. 57–71, 2014.

[26] M. Kano and Y. Nakagawa, "Data-based process monitoring, process control, and quality improvement: Recent developments and applications in steel industry," *Comput. Chem. Eng.*, vol. 32, no. 1, pp. 12–24, 2008.

[27] M. Kano, K. Nagao, S. Hasebe, I. Hashimoto, H. Ohno, R. Strauss, and B. Bakshi, "Comparison of multivariate statistical process monitoring methods with applications to the Eastman challenge problem," *Comput. Chem. Eng.*, vol. 26, no. 2, pp. 175-181, 2000.

[28] C. Shang, F. Yang, D. Huang, and W. Lyu, "Data-driven soft sensor development based on deep learning technique," *J. Process Control*, vol. 24, no. 3, pp. 223–233, 2014.

[29] G. Li, S. J. Qin, and D. H. Zhou, "Geometric properties of partial leasts quares for process monitoring," *Automatica*, vol. 46, no. 1, pp. 204–210, 2010.

[30] S. Wold, N. K. Wold, and B. Skagerberg, "Nonlinear PLS modeling," *Chemometr. Intell. Lab. Syst.*, vol. 7, no. 1-2, pp. 53–65, 1989.

[31] R. Rosipal, and L. J. Trejo, "Kernel partial least squares regression in reproducing kernel Hillbert space," *J. Mach. Learn. Res.*, vol. 2, pp. 97–123, 2001.

[32] S. X. Ding, P. Zhang, A. S. Naik, E. L. Ding, and B. Huang, "Subspace method aided data-driven design of fault detection and isolation systems," *J. Process Control* vol. 19, no. 9, pp. 1496–1510, 2009.

[33] S. X. Ding. "Data-driven design of monitoring and diagnosis systems for dynamic processes: A review of subspace technique based schemes and some recent results," *J. Process Control*, vol. 24, no. 2, pp. 431–449, 2014.

[34] M. Kano, M. Koichi, S. Hasebe, and I. Hashimoto, "Inferential control system of distillation compositions using dynamic partial least squares regression," *J. Process Control*, vol. 10, no. 2-3, pp. 157–166, 2000.

[35] K. Helland, H. E. Berntsen, and O.S. Borgen, "Recursive algorithm for partial least squares regression," *Chemom. Intell. Lab. Syst.*, vol. 14, no. 1-3, pp. 129–137, 1992.

[36] S. J. Qin, "Recursive PLS algorithms for adaptive data modeling," *Comput. Chem. Eng.*, vol. 22, no. 4-5, pp. 503–514, 1998.

[37] D. H. Zhou, G. Li, and S. J. Qin, "Total projection to latent structures for process monitoring," *AIChE J.*, vol. 56, no. 1, pp. 168–178, 2010.

[38] H. Wold, *Path models with latent variables: the NIPALS approach, Quantitative sociology: International perspectives on mathematical and statistical modeling.* New York, USA: Academic Press, 1975.

[39] A. Höskuldsson, "PLS regression methods," *J. Chemometr.*, vol. 2, no. 3, pp. 211–228, 1988.

[40] S. Yin, S. X. Ding, A. Haghani, H. Y. Hao, and P. Zhang, "A comparison study of basic data-driven fault diagnosis and process monitoring methods on the benchmark Tennessee Eastman process," *J. Process Control*, vol. 22, no. 9, pp. 1567–1581, 2012.

[41] J. Hu, C. Wen, P. Li, and T. Yuan, "Direct projection to latent variable space for fault detection," *J. Franklin I.*, vol. 351, no. 3, pp. 1226–1250, 2014.

[42] M. Basseville, *Detection of Abrupt Changes: Theory and Application*. New Jersey, USA: Prentice Hall, 1993.

[43] N. A. Adnana, I. Izadi, and T. W. Chen, "On expected detection delays for alarm systems with deadbands and delay-timers," *J. Process Control*, vol. 21, no. 9, pp. 1318–1331, 2011.

[44] N. A. Adnana, Y. Cheng, I. Izadib, and T. W. Chen, "Study of generalized delay-timers in alarm configuration," *J. Process Control*, vol. 23, no. 3, pp. 382–395, 2013.

[45] J. Xu, J. Wang, I. Izadi, and T. Chen, "Performance assessment and design for univariate alarm systems based on FAR, MAR, and AAD," *IEEE Trans. Autom. Sci. Eng.*, vol. 9, no. 2, pp. 296–307, 2012

[46] K. Zhang, H. Y. Hao, Z. W. Chen, S.X. Ding, and E.L. Ding, "Comparison study of multivariate statistics based key performance indicator monitoring approaches," in: *19th IFAC World Congress*, Cape Town, South Africa, Aug. 2014, pp. 10628–10633.

[47] S. Yang, and Q. Zhao, "Probability distribution characterisation of fault detection dealys and false alarms," *IET Control Theory A.* vol. 6, no. 7, pp. 953–961, 2012.

[48] G. Box. "Some theorems on quadratic forms applied in the study of analysis of variance problems, I. effect of inequality of variance in the one-way classification," *Ann. Math. Stat.*, vol. 25, no. 2, pp. 290–302, 1954.

[49] J. Downs, and E. Fogel, "A plant-wide industrial process control problem," *Comput. Chem. Eng.*, vol. 17, no. 3, pp. 245–255, 1993.

[50] P. Geladi, and B. R. Kowalski, "Partial least-squares regression: a tutorial," *Anal. Chim. Acta.*, vol. 185, pp. 1–17, 1986.

[51] S. Wold, "Cross validatory estimation of the number of components in factor and principal component analysis," *Technometrics*, vol. 20, pp. 397–406, 1978.

[52] H. Hotelling, "The generalization of studnts ratio," *Ann. Math. Stat.*, vol. 2, pp. 360–378, 1931.

[53] N. D. Tracy, J. C. Young, and R. L. Mason, "Multivariate control charts for individual observations," *J. Qual. Technol.* vol. 24, pp. 88–95, 1992.

[54] H. M. Wadsworth, *Handbook of Statistical Methods for Engineers and Scientists.* 2nd ed. New York, USA: McGraw-Hill, 1997.

[55] W. K. Härdle and L. Simar, *Applied Multivariate Statistical Analysis.* 3rd ed. Berlin Heidelberg, Germany: Springer-Verlag, 2012.

[56] A. E. Hoerl, and R. W. Kennard, "Ridge regression: Biased estimation for nonorthogonal problems," *Technometrics*, vol. 12, no. 1, pp. 55-67, 1970.

[57] B. C. Juricek, D. E. Seborg, and W. E. Larimore, "Identification of multivariable, linear, dynamic models: comparing regression and subspace techniques," *Ind. Eng. Chem. Res.*, vol. 41, no. 9, pp. 2185–2203, 2002.

[58] C. F. Alcala, and S. J. Qin,"Reconstruction-based contribution for process monitoring," *Automatica*, vol.45, no.7, pp. 1593–1600, 2009.

[59] C. F. Alcala and S. J. Qin, "Analysis and generalization of fault diagnosis methods for process monitoring," *J. Process Control*, vol. 21, no. 3, pp. 322–330, 2011.

[60] J. L. Liu, "Fault diagnosis using contribution plots without smearing effect on non-faulty variables," *J. Process Control*, vol. 22, no. 9, pp. 1609–1623, 2012.

[61] H. Ruben, "Probability content of regions under spherical normal distributions, IV: The distribution of homogeneous and non-Homogeneous quadratic functions of normal variables," *Ann. Math. Stat.*, vol. 33, pp. 542–570, 1962.

[62] J. Sheil, and I. O'Muircheartaigh, "The distribution of non-negative quadratic forms in normal variables," *J. R. Stat. Soc. Series C (Applied Statistics),* vol. 26, no. 1, pp. 92–98, 1977.

[63] M. Basseville, "On-board component fault detection and isolation using the statistical local approach," *Automatica*, vol. 34, no. 11, pp. 1391–1415, 1998.

[64] U. Kruger, S. Kumar, and T. Littler, "Improved principal component monitoring using the local approach," *Automatica*, vol. 43. no. 9 pp. 1532–1542, 2007.

[65] U. Kruger and G. Dimitriadis, "Diagnosis of process faults in chemical systems using a local partial least squares approach," *AIChE J.*, vol. 54, no. 10, pp. 2581–2596, 2008.

[66] H. P. Huang, C. C. Li, and J. C. Jeng, "Multiple multiplicative fault diagnosis for dynamic processes via parameter Similarity Measures," *Ind. Eng. Chem. Res.*, vol. 46, no. 13, pp. 4517–4530, 2007.

[67] J. Zeng, U. Kruger, J. Geluk, X. Wang, and L. Xie, "Detecting abnormal situations using the Kullback-Leibler divergence," *Automatica*, vol. 50, no. 11, pp. 2777–2786, 2014.

[68] J. Harmouche, C. Delpha, and D. Diallo, "Incipient fault detection and diagnosis based on KullbackLeibler divergence using Principal Component Analysis: Part I," *Singal Process.*, vol. 109, pp. 334–344, 2015.

[69] P. Phaladiganon, S. B. Kim, V. C. P. Chen, and W. Jiang, "Principal component analysis-based control charts for multivariate nonnormal distributions," *Expert Syst. Appl.*, vol. 40, no. 8, pp. 3044–3054, 2013.

[70] J. Harmouche, C. Delpha, and D. Diallo, "Incipient fault detection and diagnosis based on KullbackLeibler divergence using Principal

Component Analysis: Part II ," *Singal Process.*, vol. 94, pp. 278–287, 2015.

[71] K. Zhang, H. Y. Hao, Z. W. Chen, S. X. Ding, and K. X. Peng, "A comparison and evaluation of key performance indicator-based multivariate statistics process monitoring approaches," *J. Process Control*, vol. 33, pp. 112–126, 2015.

[72] H. Liu, Y. Tang, and H. H. Zhang, "A new chi-square approximation to the distribution of non-negative definite quadratic forms in non-central normal variables," *Comput. Stat. Data An.*, vol. 53, no. 4, pp. 853–856, 2009.

[73] J. E. Jackson and G. S. Mudholkar, "Control procedure for rsiduals associated with prinicipal component analysis," *Technometrics*, vol. 21, pp. 341–349, 1979.

[74] R. Patel and M. Toda, "Trace inequalities involving Hermitian matrices," *Linear Algebra Appl.*, vol. 23, pp. 13–20, 1979.

[75] F. Chen and Z. Zuo, "A class of inequealities on matrix norms and applications," *Gen. Math. Notes*, vol. 2, no. 1, pp. 40–44, 2011.

[76] A. M. Mathai and S. B. Provost, *Quadratic Forms in Random Variables (Statistics: A Series of Textbooks and Monographs)*. Florida, USA: CRC Press, 1992.

[77] Z. Ge, Z. Song, and F. Gao, "Review of recent research on data-based process monitoring," *Ind. Eng. Chem. Res.*, vol. 52, no. 10, pp. 3543–3562, 2013.

[78] P. Geladi, "Notes on the history and nature of partial least squares (PLS) modelling," *J. Chemometr.*, vol. 2, pp. 231–246, 1988.

[79] G. M. Morales, "Partial least squares (PLS) methods: origins, evolution and application to social sciences," *Commun. Stat.-Theor. M.*, vol. 40, no. 13, pp. 2305–2317, 2011.

[80] S. Wold, M. Sjöström and L. Eriksson, "PLS-regression: a basic tool of chemometrics," *Chemometr. Intell. Lab. Syst.*, vol. 58, no. 2, pp. 109–130, 2001.

[81] T. Kailath, A. Sayed, and B. Hassibi, *Linear Estimation.* NewJersey, USA: Prentice Hall, 1999.

[82] Z. Chen, S. X. Ding, K. Zhang, Z. Li, and Z. Hu, "Canonical correlation analysis-based fault detection methods with application to alumina evaporation process," *Control Eng. Pract.*, vol. 46, pp. 51–58, 2016.

[83] S. Lakshminarayanan, S. L. Shah, and K. Nandakumar, "Modeling and control of multivariable processes: The dynamic projection to latent structures approach," *AIChE J.*, vol. 43, no. 9, pp. 2307–2323, 1997.

[84] R. Shi and J. F. MacGregor, "Modeling of dynamic systems using latent variable and subspace methods," *J. Chemometr.*, vol. 14, no. 5-6, pp. 423–439, 2000.

[85] W. Ku, R. H. Storer, and C. Georgakis, "Disturbance detection and isolation by dynamic principal component analysis," *Chemometr. Intell. Lab. Syst.*, vol. 30, no. 1, pp. 179–196, 1995.

[86] A. Negiz and A. Cinar, "Statistical monitoring of multivariable dynamic processes with state-space models," *AIChE J.*, vol. 43, no. 8, pp. 2002–2020, 1997.

[87] R. Ergon and M. Halstensen, "Dynamic system multivariate calibration with low-sampling-rate *y* data," *J. Chemometr.*, vol. 14, no. 5-6, pp. 617–628, 2000.

[88] J. Chen and K. Liu, "On-line batch process monitoring using dynamic PCA and dynamic PLS models," *Chem. Eng. Sci.*, vol. 57, no. 1, pp. 63–75, 2002.

[89] K. Zhang, Y. A. W. Shardt, Z. Chen, S.X. Ding, and K.X. Peng, "Unit-level modelling for KPI of batch hot strip mill process using dynamic partial least squares, in: "Proceedings of the 2015 IFAC System Identification Symposium," Peking, China, Oct. 2015, pp. 1005–1010.

[90] G. Baffi, E. B. Martin, and A. J. Morris, "Non-linear dynamic projection to latent structures modelling," *Chemometr. Intell. Lab. Syst.*, vol. 52, no. 1, pp. 5–22, 2000.

[91] G. Lee, C. H. Han, and E. S. Yoon, "Multiple-fault diagnosis of the Tennessee Eastman process based on system decomposition and dynamic PLS," *Ind. Eng. Chem. Res.*, vol. 43, pp. 8037–8048, 2004.

[92] H. Wold, Estimation of principal components and related models by iterative least squares. in P.R. Krishnaiah (Editor), Multivariate Analysis, Academic Press, New York, 1966. 391–420

[93] G. Li, S. J. Qin, and D. H. Zhou, "A new method of dynamic latent variable modelling for process monitoring," *IEEE Trans. Ind. Electron.*, vol. 61, no. 11, pp. 6438–6445, 2014.

[94] H. J. Galicia, Q. P. He, and J. Wang, "A reduced order soft sensor approach and its application to a continuous digester," *J. Process Control*, vol. 21, pp. 489–500, 2011.

[95] S. J. Qin, S. Valle, and M. J. Piovoso, "On unifying multiblock analysis with application to decentralized process monitoring," *J. Chemometr.*, vol. 15, pp. 715–742, 2001.

[96] L. Ljung and E. Ljung, *System identification: Theory for the user.* Upper Saddle River, NJ: Prentice-Hall, 1999.

[97] S. J. Qin, "An overview of subspace identification," *Comput. Chem. Eng.*, vol. 30, no.10-12, pp. 1502–1513, 2006.

[98] J. Wang and S. J. Qin, "A new subspace identification approach based on principal component analysis," *J. Process Control*, vol. 12, no. 8, pp. 841–855, 2002.

[99] K. X. Peng, H. Zhong, L. Zhao, K. Xue, and Y. Ji, "Strip shape modeling and its setup strategy in hot strip mill process," *Int. J. Adv. Manuf. Technol.*, vol. 72, pp. 589–605, 2014.

[100] C. Hajiyev, "Tracy-Widom distribution based fault detection approach: Application to aircraft sensor/actuator fault detection," *ISA Trans.*, vol. 51, pp. 189–197, 2012.

[101] E. Saccentia, A. K. Smilde, J. A. Westerhuis, and M. W. B. Hendriks, "TracyWidom statistic for the largest eigenvalue of autoscaled real matrices," *J. Chemometr.*, vol. 25, no. 12, pp. 644–652, 2011.

[102] L. Xie, J. Zeng, U. Kruger, X. Wang, and J. Geluk, "Fault detection in dynamic systems using the Kullback-Leibler divergence," *Control Eng. Pract.*, vol. 43, pp. 39–48, 2015.

[103] L. H. Chiang and R. D. Braatz, "Process monitoring using causal map and multivariate statistics: Fault detection and identification," *Chemom. Intell. Lab. Syst.*, vol. 65, no. 2, pp. 159–178, 2003.

[104] C. A. Tracy and H. Widom, "The distribution of the largest eigenvalue in the Gaussian ensembles. In: van Diejen J, Vinet L, editors. Calogero-moser-sutherland models," New York: Springer; 2000. pp. 461–472.

[105] S. Kourouklis and P. G. Moschopoulos, "On distribution of the trace of a noncentral Wishart matrix," *Metron XLIII*, vol. (1-2), pp. 85-92, 1985.

[106] N. R. Goodman, "The distribution of the determinant of a complex Wishart distributed matrix," *Ann. Math. Stat.*, vol. 34, no.1, 178–180, 1963.

[107] B. Arthur, D. K. J. Lin, and R. N. McGrath, "Multivariate control charts for monitoring covariance matrix: A Review," *Qual. Technol. Quant. M.*, vol. 3, no. 4, pp. 415–436, 2006.

[108] J. Guerrero, "Multivariate mutual information sampling distribution with applications," *Commun. Stat. Theory Methods*, vol. 23, no. 5, pp. 1319–1339, 1994.

[109] C. R. Rao, *Linear Statistical Inference and its Applications.* 2nd ed. New Jersey, USA: Wiley, 2008.

[110] S. Yin, H. Luo, and S. X. Ding, "Real-time implementation of fault-tolerant control systems with performance optimization," *IEEE Trans. Ind. Electron.*, vol. 64, no. 5, pp. 2402–2411, 2014.

[111] S. Yin, S. X. Ding, X. Xie, and H. Luo, "A review on basic data-driven approaches for industrial process monitoring," *IEEE Trans. Ind. Electron.*, vol. 61, no. 11, pp. 6418–6428, 2014.

[112] S. Yin, *Data-driven design of fault diagnosis systems.* Düsseldorf, Germany: VDI Verlag, 2012.

[113] S. Yin, G. Wang, and H. Karimi, "Data-driven design of robust fault detection system for wind turbines," *Mechatronics*, vol. 24, no. 4, pp. 298–306, 2014.

[114] H. Hao, "Key performance Monitoring and Dignaoisis in Industrial Automation Processes," Ph.D Thesis, 2014.

[115] V. Venkatasubramanian, R. Rengaswamy, and S. N. Kavuri, "A review of process fault detection and diagnosis Part I: Quantitative model-based methods," *Comput. Chem. Eng.*, vol. 27, no. 3, pp. 93–311, 2003.

[116] V. Venkatasubramanian, R. Rengaswamy, and S. N. Kavuri, "A review of process fault detection and diagnosis Part II: Qualitative models and search strategies," *Comput. Chem. Eng.*, vol. 27, no. 3, pp. 313–326, 2003.

[117] V. Venkatasubramanian, R. Rengaswamy, and S. N. Kavuri, "A review of process fault detection and diagnosis Part III: Process history based methods," *Comput. Chem. Eng.*, vol. 27, no. 3, pp. 327–346, 2003.

[118] Y. A. W. Shardt, Y. Zhao, F. Qi, K. Lee, X. Yu, B. Huang, and S. Shah, "Determining the state of a process control system: current trends and future challenges," *Can. J. Chem. Eng.*, vol. 90, 217–245, 2012.

[119] A. Haghani, T. Jeinsch, and S. X. Ding, "Quality-related fault detection in industrial multimode dynamic processes," *IEEE Trans. Ind. Electron.* vol. 61, no. 11, pp. 6446–6453, 2014.

[120] Z. Q. Ge and Z.H. Song, "Multimode process monitoring based on Bayesian method," *J. Chemometr.*, vol. 23, no. 12, pp. 636–650, 2009.

[121] H. Wold, "Nonlinear iterative partial least squares (NIPALS) modeling: some current developments, in P.R. Krishnaiah ed. Multivariate Analysis II," in: *Proceedings of an International Symposium on Multivariate Analysis*, Dayton, Ohio, USA, June, 1972, pp. 383–407.

[122] B. S. Dayal, and J. F. MacGregor, "Improved PLS algorithms," *J. Chemometr.*, vol. 11, no. 1, pp. 73–85, 1997.

[123] S. de Jong, "SIMPLS: an alternative appraoch to partial least squares regression," *Chemometr. Intell. Lab. Syst.*, vol. 18, pp. 251–263, 1993.

[124] A. S. Naik, S. Yin, S. X. Ding, and P. Zhang, "Recursive identification algorithms to design fault detection systems," *J. Process Control*, vol. 20, no. 8, pp. 957–965, 2010.

[125] L. M. Elshenawy, S. Yin, A. S. Naik, and S. X. Ding, "Efficient recursive principal component analysis algorithms for process monitoring," *Ind. Eng. Chem. Res.*, vol. 49, no. 1, pp. 252–259, 2010.

[126] B. A. Altaf, "Application of dynamic partial least squares to Complex Processes," PhD Thesis, 2013.

[127] P. A. Hassel, "Nonlinear partial least squares," PhD Thesis, 2003.

[128] K. Zhang, Y. A. W. Shardt, Z. Chen, S. X. Ding, and K. Peng, "Using the expected detection delay to assess the performance of different multivariate statistics process monitoring methods for multiplicative and drift faults," Submitted to *J. Process Control*.

[129] K. Zhang, Y. A. W. Shardt, Z. Chen, S. X. Ding, and K. Peng,"A comparison of T^2- and Q-statistics for detecting additive and independent multiplicative faults," Submitted to *J. Frankli. I.*.

[130] Y. A. W. Shardt, *Statistics for Chemical and Process Engineers: A Modern Approach.* Cham, Switzerland: Springer, 2015.

[131] Y. Zheng, N. Li, and S. Li, "Hot-rolled strip laminar cooling process plant-wide temperature monitoring and control, " *Control Eng. Pract.*, vol. 21, no. 1, pp. 23–30, 2013.

[132] C. M. Bishop, *Pattern Recognition and Machine Learning.* Cambridge, U. K: Springer, 2006.

[133] C. Aldrich, L. Auret, *Unsupervised Process Monitoring and Fault Diagnosis with Machine Learning Methods.* London, U. K: Springer, 2013.

[134] D. Gorinevsky, "Fault isolation in data-driven multivariate process monitoring," *IEEE Trans. Control Syst. Technol.*, vol. 23, no. 5, pp. 1840–1852, 2015.

Printed in the United States
By Bookmasters